DATE DUE

SCIENCE AND CRITICISM

SCIENCE AND CRITICISM

THE HUMANISTIC TRADITION IN CONTEMPORARY THOUGHT

BY

HERBERT J. MULLER

NEW HAVEN AND LONDON
YALE UNIVERSITY PRESS

Third printing, August 1964.
Printed in the United States of America by
The Carl Purington Rollins Printing-Office of the
Yale University Press, New Haven, Connecticut.

Originally published as a Dwight Harrington Terry
Foundation Lecture.

To Janet

PREFACE TO THE FIRST EDITION

I BEGAN this work with the general aim of consolidating and extending the traditional position of humanism, as represented most conspicuously and honorably in literature today by Thomas Mann. In the *Sturm und Drang* of the modern world, Mann has sought to preserve "the idea of supra-personal, supra-partisan, supra-racial standards and values"; he has aspired to a harmonious synthesis—of nature and culture, of tradition and innovation—in which apparent opposites are resolved by assimilation rather than annihilation; and to him this gentlemanly attitude is nevertheless "revolutionary" because it refuses to accept any tradition uncritically. More specifically, however, I wished to correlate this attitude with scientific thought, which is now the chief source of innovation. In the end I found myself devoted mainly to an effort to make really available, for the purposes of literary criticism, the revolutionary findings in the natural and social sciences, with which critics are generally familiar but of which they generally make only superficial, incidental, or erratic use.

This is scarcely a modest aim. Yet I have not tried to offer a tight formal system or a certified solution: we already have too many closed systems of thought, and humanism and science are alike in not pretending to give a final answer. Neither have I tried to write an encyclopedia of modern science: the field is very large and I am not H. G. Wells. My effort has been simply to make a beginning: to inquire into first principles, to work out some implications of some of the more significant developments, to point to some ways and means of orientation—at most to write a kind of grammar and glossary. In Locke's words, my study is "fitted to men of my own size." Nor am I attempting to disarm criticism when I add that it is fitted in particular to men of my own time; for in these days critics are disarmed about as easily as dictators.

Nevertheless it remains a broad study, and calls for further qualification. As the layman has to get all his science at second hand, he cannot be sure of the soundness of any theory or of his choice between conflicting theories; he can be sure only that his work will seem as superficial to the specialist as abstruse to other laymen. Furthermore, an exposition of science is necessarily a selection, a drawing out of implications is an interpretation, and both are naturally influenced by special interests. Because the critic is looking *for* something, he is likely to board the theory that will take him where he wants to go. My own primary interest is literary humanism; it is a catholic disposition but still a predisposition, or in other words a bias. Similarly, the effort to cover so much ground involves the risks of scatter and sprawl, with a final collapse into mere miscellany. In following out various leads, I have consciously sacrificed neatness of design and severity of logic for the sake of possible breadth and suggestiveness; I have not worried too much over loose ends. No doubt, too, I have fallen into some inconsistencies. Words are so chameleonlike that they are bound to take on different meanings in the course of a protracted argument, especially as the author himself discovers and learns during its course, and here I have deliberately tried to multiply meanings, deliberately translated the exact language of science into literary and often colloquial terms. Nevertheless I believe that my work is governed by a consistent attitude, that its approaches are toward a consistent goal, that its marginalia fall along consistent lines of direction—that it is a fusion and not a confusion. It must bear criticism accordingly.

In all this I go far beyond specifically esthetic problems. I do not make this statement proudly, however. Contemporaries are apt to be somewhat scornful or fearful of the word "esthetic," as if it were a synonym for genteel, trivial, irrelevant, irresponsible. Actually, the esthetic analysis is the peculiar business of the literary critic, should be his first and last operation; and it is indeed the most difficult to perform. At the same time, I do not apologize for my own concern here. Esthetic interests should be distinguished from other vital interests but cannot be divorced, least of all by one who values them; some philosophy is implicated in any analysis or judgment, and a philosophy of art is ultimately a philosophy of life. For that matter, the critic is constantly referring to general ideas even during his specialized job, often pausing to explain or argue such ideas, often retracing his steps to clarify them for himself. Like it or not, he has some kind of philosophy, and he might better have a conscious and coherent one.

The important thing is to call things by their right names—and not to call the other fellow names because he is interested in different things. Such sniping is the more trivial today, for all literary fellows are now in the same boat. The world is too much with them, or against them; as they try to do their own work while the very existence of their world is threatened, all have to contend with the same feeling of irrelevance, irresponsibility, impotence. In these days it is hard to think as a good humanist, and harder to feel like one. The venerable tradition seems ancient indeed, far away and long ago; often enough I am tempted to the ultimate futility of believing in Man and despairing of men. Yet one must work on the assumption that all thinkers have always worked on, affirmed in the mere act of writing even if they wrote to deny it—the assumption that thought is worth while. If civilization is destroyed, the soldier and the businessman will go down with the poet and the critic, and all distinctions between academic and practical will be lost in a universal irrelevance. If it survives, these literary interests will not be academic. My faith remains that it is worth saving primarily because of the values and ideals of humanism.

Meanwhile one at least need not fear that an effort at comprehensive survey and disinterested criticism will do any harm to speak of. He can be sure that he will not distract too many readers from the urgent business of the world.

PREFACE TO THE SECOND EDITION

ALMOST fifteen years have passed since I finished this book; and at the rate that history is now made, it may seem fifty years ago. D-Day, VE-Day, and VJ-Day belong to another generation; Roosevelt and Stalin have followed Mussolini and Hitler into the shades. We have seen the establishment of the United Nations, a continuous cold war, another shooting war in Korea. Science has given us atomic power, and with it the problem of how to live with hydrogen bombs—not to mention rockets, planes with supersonic speeds, and projects for artificial satellites. In everyday life we contemplate the possibilities of automation. On the frontiers of scientific thought, beyond the ken of laymen still trying to catch up with the implications of relativity, Einstein offered a new comprehensive equation. Other developments include Hoyle's theory of the mysterious universe, which makes it seem still more fantastic. In all fields there have been more or less significant developments, from the "new criticism" in literature to the "new conservatism" in political thought. It is characteristic of our time that even conservatism must be "new."

Yet it is still the same old revolutionary world, facing basically the same problems, engaged in the same race between education and catastrophe. I was gratified, and a little depressed, to find that this book does not seem musty or outmoded to me. While I cannot pretend to have kept up with all the developments in science, I believe they have not invalidated my account of its basic concepts and their philosophical implications. Were I to rewrite the book I would naturally make some additions and revisions (and incidentally try to iron out a cryptic quality in the style which sometimes suggests that I mistook myself for Ralph Waldo Emerson); but I would make no fundamental changes. My humanistic position on literature, philosophy, religion, democracy, and values in general likewise remains substantially the same.

I have learned something, however, from further study and reflection, as well as the experience of being reviewed. When the book first appeared, a kindly philosopher wrote me that it was one that every professor of English ought to read, and that none would. To my knowledge few of them have read it; it has apparently aroused more interest among scientists. In fairness to my colleagues—and to new readers—I should say that the title of the book is misleading. Although my declared intention was to make the revolutionary findings in modern science available "for the purposes of literary criticism," and these purposes in fact got me started, I was led far afield. Some of the implications I conscientiously drew for critics were dragged back by the horns. The book is really directed to students of the humanities, not primarily to students of literature. But even so I must still believe that its subject matter is a proper concern for literary men, especially those who deplore the narrowness or inhumanity of science. Many are given to generalizations about modern culture that imply a knowledge of history, philosophy, and science they do not actually have.

Another source of possible misunderstanding is the problem of emphasis or timing that must always plague writers who, like me, are given to the depressing habit of thinking "on the other hand." In the section on Freud, for instance, I begin and end with what I thought was a clear emphatic statement of the great importance and value of his work, but I dwell chiefly on its limitations as an exact science, and on the common abuses of it; at the time this seemed to me what most needed to be said. As a result I got letters thanking me for having annihilated Freud, while one distinguished reviewer declared that at this point he lost all confidence in the book—though to my mind he had acknowledged much the same limitations and dangers in a more reverential essay on Freud and literature. There is no avoiding such misunderstanding, I suppose. Still, the fact remains that I devote two pages to praise of Freud's achievement, twelve pages to criticism of it. Today my opinion of it is essentially the same, but I would expand those two pages. Similarly I might shift my emphasis somewhat in other sections. As it is, I can only hope that readers will heed both what goes before and what comes after "the other hand."

The chief qualification I should now add, finally, is one I have already indicated in *The Uses of the Past.* In some of my generalizations about human nature I reflected the common provincialism or conceit of the West, describing as the needs of "man" what are at most the needs of Western man, or more strictly of some Western men. Thus in the section on the naturalistic basis of values I asserted that men "naturally aspire to abundance of life, fullness of being"; whereas all history indicates that most men have aspired to no such thing, much less to the freedom that is the necessary condition of abundant life. In the East men have "naturally" taken to the ways of passive acceptance, obedience, resignation, or even renunciation. Most of the moral and religious teachers of mankind have counseled these ways; until recent centuries very few thinkers have encouraged the aspiration to freedom and abundance. The liberal, humanistic position is a rare one in world history. It is strictly a faith.

Nevertheless I still hold to this faith. I should repeat that it is at least a natural expression of the distinctive potentialities emerging from the biological and social evolution of man. It therefore has a natural appeal to men once they become alive to these possibilities; the East is now astir with the Western spirit, seeking more of the goods of this life instead of a compensatory life to come. And though this means more possibilities of conflict and disorder, I still see no better hope for a decent future than the liberal faith.

November, 1955

ACKNOWLEDGMENTS

SOME of this material has already been printed in various periodicals. Specifically, section 5 of Chapter II is based on "The Functions of Criticism," in the *Saturday Review of Literature;* section 6 of Chapter II on "Exercises in Incongruity," in the *Yale Review;* the first three sections of Chapter IV on "Humanism in the World of Einstein," in the *Southern Review;* section 3 of Chapter VII on "Pareto, Right and Wrong," in the *Virginia Quarterly Review;* and section 2 of Chapter VIII on "Scientist and Man of Letters" in the *Yale Review*. The opening chapter also contains a few short passages from "Recent Pathways in Criticism" and "Matthew Arnold: A Parable for Partisans," both in the *Southern Review,* and from "Literary Criticism: Cudgel or Scales?" in the *American Scholar*. All these articles, however, have been broken up, condensed, or expanded, and in general completely revised to fit into the scheme of this book.

To various publishers I am indebted for permission to quote: to the Macmillan Company for excerpts from *The Logic of Modern Psychology,* by Carroll C. Pratt, and from *Anatomy and the Problem of Behavior,* by G. E. Coghill (published by the Cambridge University Press, England); to Houghton Mifflin Company for excerpts from *Patterns of Culture,* by Ruth Benedict; to Harcourt, Brace and Company, Inc., for excerpts from "East Coker," by T. S. Eliot; and to the Open Court Publishing Company for excerpts from *Experience and Nature* by John Dewey.

I am very grateful to my friends and colleagues, Professors Herbert L. Creek, Emerson G. Sutcliffe, and John R. Lindsay, and to Professor Charles W. Hendel of Yale University, for helpful suggestions and for criticism of my manuscript.

I also wish to express my deep gratitude to the John Simon Guggenheim Memorial Foundation, whose generosity assisted me in embarking upon this work.

CONTENTS

PART I

THE THEME: PROBLEMS AND PREMISES

SCIENCE AND CRITICISM

I

THE LITERARY CRITIC IN THE CONTEMPORARY WORLD

By reason of this attainment of self-consciousness by the will for truth, morality from henceforward—there is no doubt about it—goes *to pieces:* this is that great hundred-act play that is reserved for the next two centuries of Europe, the most terrible, the most mysterious, and perhaps also the most hopeful of all plays. NIETZSCHE.

1. THE PROBLEM

ALTHOUGH all ages are ages of transition, because all lead to the next age, the processes of transition were never more rapid, violent, and conspicuous than now. Nevertheless they also seem to be obscure. As we run we may read oddly contradictory accounts of whence we have come, where we are headed, and why we are running. The trouble with us, it appears, is that we have too much science, or too little; we are too complacent about our intellectual emancipation, or we have too little faith in the power of intelligence; we have stifled natural instinct and emotion, or we are at the mercy of instinct and emotion; we are cut off from our past, or we need to kill more dead men. We are cursed by profound uncertainty and inordinate pride, standardization and lack of standards, regimentation and restlessness, heritage and heresy. Similarly our literature suffers from escapism and from propaganda, a poverty of ideas and an excess of ideologies, a cult of unintelligibility and a persistent vulgarization, an exclusive cultivation of private sensibility and a smothering of the individual under social documents, a supersubtle, highly mannered school and a hard-boiled, ill-mannered school. And a good case can be made out for any of these indictments of modern life and literature. In other words, this is in fact a very complex society.

Now these incongruities are significant, they must be taken into account. If they make impossible a tidy account of the modern world, they suggest an approach through the principle of incongruity itself. They are obvious signs of disorder, disharmony, discontinuity, disintegration—the profound disruption of the inner unity of our experience underlying the immense confusion of its outer forms. They are also signs of power and wealth, the many kinds it indeed takes to make a world, especially a civilized one. And the central source of all the contradiction, confusion, and abundance is plainly science. It has long since revolutionized habits of thought, modes of living, the material means of life; it has not yet readapted habits of feeling, basic patterns of living, the spiritual ends of life. Underlying all the immediate issues in the realm of affairs as in the realm of thought is the basic need of orientation and integration. The problem, generally, is to harmonize our very old interests with our new conceptions of the world and new conditions of life. It is to unite natural knowledge and natural piety, reason and ritual, "science and sanctity." It is actually to pull ourselves together.

More serious than the notorious "conflicts," between science and literature or religion and business, is the *separation* of these major interests. Cross-purposes do cross, and so may conceivably be reconciled; isolate purposes may never meet at all. A specialization in the interests of freedom and efficiency has in fact led to bureaucratic narrowness and waste. Hence John Crowe Ransom describes this godless, loose-laced age as "Puritan." Religion becomes Protestant, divorced from ritual and art as from the state; business preaches the gospel of *laissez faire;* art also aspires to autonomy, in the name of "purity"; science becomes positivist and dismisses as "meaningless" the vital questions it cannot answer. All resent interference by other interests, resist the demand for integration. Keep the government out of business, keep religion out of public affairs, keep public affairs out of science, or art, or education—"Keep everything out of everything else," writes Kenneth Burke, "and then complain that things are in pieces." "Gross materialism" is the common label for our condition, but "pure idealism" would stick as well. Religion, philosophy, science, or art for its own sake is finally as narrow and unillumined as business for its own sake. They are alike monopolists of capital, to use John Dewey's words: "And the monopolization of spiritual capital may in the end be more harmful than that of material capital."

These descriptions are partial; there is also a strong centralizing tendency. But they are less arbitrary and exclusive than more popular descriptions. In viewing the modern world simply as a wasteland, for example, one does not attack the problem but gives in to it. One does not even locate it. An age that has displayed such abundant energy and enterprise, that has produced so much bold pioneering achievement, that has so enormously increased the powers and extended the horizon of man, that feels itself capable of almost any possibility and at its most fearful is fearful of its own power and strength—such an age cannot be considered merely decadent or God-forsaken. Say the worst and it is not feebly confused but terribly perverse, it is not effete but has bred its own barbarian hordes. As shallow, in literary criticism, is the view Mary Colum expressed in *From These Roots:* a view of a great cultural depression in which critics everywhere are intellectually and spiritually impoverished, unemployed, underprincipled. In fact there is a remarkable busyness, a wealth of ambitious, resourceful, original, even brilliant work. Edmund Wilson, Kenneth Burke, John Crowe Ransom, Allen Tate, R. P. Blackmur, T. S. Eliot, I. A. Richards, William Empson—these are but a few names; and I should say that for subtle, acute analysis they cannot be matched in any earlier period. The analysis is often a diagnosis, the very activity is an avowal of our especial need of taking stock. But the fact remains that this is an age of criticism, and may well go down as one of the greatest.

All too resolute, on the other hand, are the various militarist solutions of our problem. We have surrendered, as Santayana observed of the Renaissance, "the essence for the miscellany of human power." Men are accordingly seeking essences. They are accordingly prone to seize on a single principle, to achieve unity by simple mandate, finally to set up some absolutism that is the philosophical equivalent of the totalitarian state. And they are as ruthless as the dictators in their suppression of major interests and hard-earned goods. Integration, once more, is the key to our problem; but it is not a simple key. Morons, beauty queens, salesmen, prize fighters, politicians, racketeers —all are likely to be better integrated than philosophers or poets; nor in this respect can our whole society hope to surpass most primitive societies, much less any anthill. The important question is always *what* has been integrated and at what expense, how much included and how much sacrificed.

In general, the tendency is not to reconcile but simply to eliminate. Men of science, men given to "realism," are likely to make a clean sweep of old interests and sentiments as so much rubbish. They regard religion as superstition, metaphysics as moonshine, art as primitive pastime, and all ritual as monkey-business; they apparently assume that because the traditional answers to old needs have been discredited, the needs themselves have vanished. But literary men as often try to ignore or suppress the necessities of the present. When they do not attack science—or even intellect—as sheer evil, they seek anxiously to minimize it, keep it out of the "world of spirit," restrict it to the "material" world—a kind of red-light district where it is legalized only so that it may be regulated. The spiritual children of Babbitt and More (still alive in the universities) set up standards no less rigid for having no perceptible foundation in heaven or on earth; T. S. Eliot clings with mournful fidelity to a sovereign orthodoxy that apparently exists by itself, independently of men and their doings; others simply pack up and take off on the ancient flight "from the alone to the Alone." Similarly the University of Chicago bravely announces the need of making education a unified discipline instead of an elegant smattering—only to elect the medieval scholastic discipline, a scheme whose content is hopelessly inadequate to the demands of modern knowledge and experience. In short, positivists and pietists alike fail to size up the whole situation and fight their way through it. They begin by ruling out the opposition, they end by intensifying the actual problem. They deepen and widen the split in the modern world, because of which knowledge is too often undisciplined power and piety is too often mere posture.

Altogether, modern thought is much like modern industry in its production at once of amazing wealth and shocking waste. For literary critics specifically there is now available more stimulating thought and exciting possibility than ever before. They have learned much about the conditions out of which literature grows, they can see it more clearly in its organic relation to our whole culture, they can give more substance to the swelling adjectives that describe its effects. They have as well the advantage implicit in the wholesale revaluation that is their especial problem: everywhere men are busy airing out first principles, overhauling universal laws, dusting off eternal verities, cleaning the windows of the mind.

Yet no less plain is the neglect or the abuse of these rich resources. Men tend to exploit them with no regard for the whole intellectual economy; they invest all their intellectual capital in some one idea and then market a standard brand of truth; their final aim is monopoly, a corner in truth. I employ the economic metaphor advisedly. In a capitalistic society, critics too are prone to regard their work as a competitive rather than a coöperative venture; they too are prone to become specialists or efficiency experts, and aspire to an air-tight perfection. Because so much learning is available, moreover, a little of it is more dangerous than ever before. All about are exciting ideas. Men are bound to run into one, and then they are likely (to borrow a figure from William James) to lie down on it with their whole weight, like a cow on a doorstep, so that one can get neither in nor out with them. But behind all such abuses lies the schooling of two thousand years, the holy quest of absolute truth, and for all seekers of the Truth the promise remains: Seek and ye shall find. Despite their scientific or sophisticated manner, men have the same hankering as their pious ancestors for a cozy universe, a closed system of certainties erected upon a single principle. Hence all the new ways of looking at things have not so much discouraged the passion to say the last word as suggested new ways of saying it. And hence, though rumor has it that this is an age of immense complexity and profound uncertainty, in fact one is always hearing utter simplicities announced with the utmost positiveness.

This immemorial demand of thinkers, that all existence bow to their requirements, is brave. It is also a little insolent. When Hitler makes such demands he is called a monster. And if, in Hitler's world, violently simplified attitudes are more understandable, up to a point necessary, they are more dangerous as well. The serious objection to them is not merely on abstract grounds of philosophical wisdom. It is in the end profoundly practical. Whether radical or reactionary, the doctrinaires make for less efficient dealings with a shifting, complex, recalcitrant world—a world notoriously indifferent to man's desire for simplicity and security. They make for an arbitrary fixation of attitude before possibilities have been explored or even recognized. They make for crude, limited, automatic responses to verbal symbols, to the noise of "isms," when a sensitive, flexible, resourceful adaptation to objective facts is especially needful. In the end, they condition inappropriate behavior precisely as does the Hollywood way of looking at life, and as surely they tend to leave men unprepared for all the natural shocks that flesh is heir to.

Here, then, is the argument for a comprehensive, dispassionate survey of our possibilities, an effort to order the wealth of thought confused and obscured by narrow specialization, rigid dogma, and violent negation. More specifically, here is the argument for a pragmatic, humanistic approach. It is the most catholic and flexible philosophy, the most hospitable to the claims of both past and present, and therefore the most suitable for purposes of orientation and integration. It can enable a vital, harmonious synthesis, the widest possible comprehension of experience with the least possible sacrifice of major interests and values. It is the best adapted to a world on the make simply because it is always on principle an *approach.* For in the face of the multiple facts of experience, of the history of human thought, of the endless warfare among the soldiers of eternity, any serious pretension to finality is preposterous arrogance.

2. THE APPROACH: DISCOUNTS

As the obvious is of all things the most difficult to keep in mind, especially in an age given to subtlety, let us rehearse some platitudes. Experience still has manifold aspects and mysterious depths, truth is never whole and never simple, some element of truth gives vitality to every living fallacy. All theories are disputable and exist for purposes of dispute, all dogmas are both pertinent and impertinent; all correspond in some way to human needs and purposes, none corresponds perfectly to Reality. What a man sees depends not only on the keenness of his sight but on where he stands and what he looks for. What a mature man sees is in any event very mixed. The Primal Curse, E. M. Forster has remarked, is not the knowledge of good and evil but the knowledge of good-and-evil; and so it is. Yet it is also the ultimate wisdom, and at any rate the inescapable price paid by the sons of Adam.

As plainly, however, men always seek the absolute, the reality behind the appearance, and often act *as if* they had got hold of it. They cannot do otherwise; experience itself is an endless assertion of choice. In the very act of "having an idea," moreover, they necessarily force or exaggerate the idea, for to become clearly conscious of something is to isolate it from its context, give it an independence and prominence that are strictly untrue. When they talk or write, they must appear to be even brasher, for they cannot possibly hedge every remark with all the necessary qualifications. But they can return to good sense through the "as if." These words are a clue to the logic of science, the cornerstone of an elaborate philosophy; their implications can fill volumes. For the moment it is enough to keep them at least in the back of one's head, recalling Samuel Johnson's advice on the cant proper to polite discourse: "You may *talk* in this manner; it is a mode of talking in Society: but don't *think* foolishly."

Likewise the thinker will do well, before and after every job, to take a good look at everyday experience, practical activity, common-sense solutions. In this way he can escape many artificial problems and keep the real ones life-sized. An example (to which I shall keep returning) is all the anguish that has been stirred by the idea of determinism. Science, we hear, has made it logically impossible for us to hang on to the principle of free will, without which man can have no responsibility and his life no dignity. As a matter of fact, science—as science—does not even attempt to settle the issue. But meanwhile it has plainly increased our actual freedom. In theory, its laws may be depressing; in practice, the more of them we discover the more things we can do to our world. At the same time, all our reasoning and all our behavior acknowledge the principle of necessity. The man who is most convinced that he is captain of his soul is also sure that he can understand and explain the behavior of his neighbor—as he could not were it lawless; despite our intuitive conviction of freedom and our passion for more, we would be far more miserable if we could not count on necessities. There remains a real issue for philosophers, and a practical problem for scientists. All serious thought is finally an effort to know the nature of our necessities, the limits of our freedom. But thinkers too often forget that in our practical activity we assume both determinism and free determination, and can live intelligently and comfortably without deciding the matter once and for all.

A systematic discounting on the basis of such platitudes and common-sense observations can be, I believe, the most effective of preliminary operations. Alike, the many helpful ways of interpreting human experience play up certain vital interests and suppress others. Hence rival philosophies—idealism and materialism, monism and dualism, classicism and romanticism—have tumbled down the centuries, "like two boys playing leap frog," observes E. T. Bell. Each reaction satisfies neglected needs and emphasizes neglected truths but in turn ignores or denies others. The two-party system works to good purpose, in philosophy as in politics; all the major interests get represented and in time are sure to have their turn in power. Like political parties, however, systems of thought tend to supplant rather than to supplement. They become as rigid or reactionary as vested interests, their followers vote a straight ticket. And today, when so many obviously useful systems are competing—positivist, behaviorist, Marxist, Freudian, Gestaltist, etc.—simple realism, not to say simple thriftiness, calls for a moratorium on simple certainties. We cannot make the best use of any one until we have discounted the claims of all, recognized that no one is sufficient.

To think is necessarily to simplify and abstract; only so can one order experience and give it significance. But the thinker should be aware that he is simplifying, and that all abstractions are metaphors or crude approximations. To think is also to pursue certain objectives, inevitably conditioned by the *Zeitgeist,* immediate needs, and individual temperament; the true title of every book, it has been said, is *How to Be More Like Me.* But the thinker should still strive to get outside his age and see around his purposes. He should use his concepts as sights or leads, not as fixed, necessary principles. He should regard the whole scheme of his thought as a practical means of dealing with reality and not a literal transcript of reality, a vehicle and not a destination. He should remember that thought is indeed a speculation, a gamble. The one clear certainty that emerges from the history of human thought is that a perspective is in fact a perspective, not a panorama of the universe, and that the drama of this history is always presented from the limited point of view of an actor in it, not of the omniscient author. In a word, one should be content with truths instead of the Truth.

This is not to say that one should never commit himself, and never, never judge. If all principles were accepted at par, none would have worth. All call for criticism, some necessarily for opposition. One who cherishes the ideal of tolerance may enfold Fascists in the mantle of compassionate understanding, conscientiously remembering that they too are God's chillun; in the very name of this ideal he has nevertheless to take a stand against them. Or a better illustration, because calling for finer discrimination, is the problem of mysticism. In a scientific age, the critic is likely to be unsympathetic to mystical experience. Yet on the face of it such experience has great value for the individual. It is possible, too, that the mystic really is in touch with an Oversoul, for his claims cannot be absolutely disproved. At the same time, neither can they be proved. Mystical experience does not demand a supernatural explanation; the critic may say that it differs only in intensity, not in kind, from ordinary esthetic experience. He may add that it has a limited social value, simply because it is so utterly private and ineffable, and that if it were followed through and universalized in a great society of mystics, the result would be a chaos beyond Milton's powers of description. Above all, however, he has to combat the customary rationalizations of this experience: the exaltation of the unconscious, the opposition to intellect, the disparagement, generally, of rational, controlled, communal purposes that generations of men have as undeniably found valuable. In short, the cue is assimilation by discount. It is to accept the positive contributions and welcome the possibilities, to combat the blighting negations and correct the distortions.

Now I have called this approach "humanistic." It is not a precise term, it appears in very mixed and sometimes dubious company, and for logical purposes it has the further disadvantage of evoking easy, agreeable emotion. Yet it is adequate for my present purposes, which are not rigorously logical. I am recommending a general attitude, a state of mind rather than a system of thought; my very point is that the term must not keep exclusive company. Even the strategy of the emotional appeal—the connotation of "humanity," the implication that opponents are "antihuman"—is not simply unscrupulous. The humanism of tradition evokes a reverent emotion because it is so conspicuously *humane* an attitude. And it is conspicuous because the dogmatic attitudes that men consider natural, very "human," are nevertheless essentially inhuman. Thinkers are forever trying to shirk their share in the Primal Curse.

The fundamental condition of value, Ramon Fernandez points out, is not only a clear perception of the realities one combats but a recognition of the *dignity* of these realities. Reformers of all stamps are prone to regard the existing order as sheer folly or evil, and its resistance as purely external—the resistance of a wall to being knocked down. Hence the tragic sense is replaced by a brutal sensation of catastrophe, hence the event often becomes catastrophe. Even the sacrifices of these reformers actually cost them too little effort. Today such contempt of the opposition is made still easier by the suspicion of all professed motives, the disbelief in the purity or disinterestedness of any idealism. Contemporaries have learned from Marx that ideals are masks for selfish class interests or refuges from responsibility; from Freud that they are wish fulfillments or offspring of the blind monsters in the unconscious; from behaviorism that they are conditioned reflexes comparable to the salivation of a laboratory dog; from semantics that they are meaningless noises. Most men, to be sure, have debunked only the other fellow's motives and values: the bourgeois intellectuals of Marxism are themselves above class prejudice, Freudians have attuned their wishes to the real reality, behaviorists are conditioned to the right ideals, semanticists make only meaningful noises, and all are "scientific," disinterested seekers of pure truth. But the other fellow repays in kind, each species of analyst can deflate all the others, all values go down together. The result is a widespread demoralization, appearing on the lower intellectual levels as a shallow or professional cynicism, on the higher levels as a terrible sincerity that makes it impossible to be sincere. Men of good will are haunted by a fear of being unrealistic, their most honorable scruples are exploited; an inhuman arrogance calls out an unnatural humility, in which intelligence disbelieves in its natural reason for being.

We therefore need again to return to the obvious. Idealism does not end where it begins, that which conditions it is not its cause or its essence, mixed stuff is in fact mixed. That the causes of important events are always less admirable than the sentiments accompanying them does not mean that the sentiments are an idle sham. Plainly they are real, they have force, they have worth; no thinking is more wishful than that which ignores them or tries to explain them away. And the humanist can make good use of all the new insights into the sources and conditions of human values because he also insists on the necessity and the dignity of these values. He can more nearly realize the ideal of the debunker himself, the aim to "tell all." In short, there is practical wisdom as well as piety in Bacon's prayer: "May we not be wise above measure and sobriety, but cultivate truth in charity."

This attitude has been symbolized by such names as Chaucer, Erasmus, Rabelais, and Montaigne, and in more recent times by Matthew Arnold, William James, Thomas Mann. To call the roll is to acknowledge that it is indeed a very general attitude. Nevertheless this diverse company has given it vitality and substance; and this study, once more, is based on the assumption that today it can be revitalized and made still more substantial by a correlation with scientific thought. For the moment, I conclude with the restatement of it by Kenneth Burke. Contemporaries are likely to be embarrassed by Arnold's vocabulary, suspicious of James's heartiness and his willingness to try God, simply awed by the Olympian qualities of Mann; but Burke is as subtle, sophisticated, and altogether modern worldly as they could wish. In *Attitudes toward History,* he sets out from the premise that a cultural frame is never broad enough to embrace all the necessary interests, never flexible enough to adapt itself readily to changing needs; a society inevitably overstresses certain attitudes and understresses others. Proper discounting, however, can make all thought usable and is the first step toward higher synthesis. Burke accordingly urges "comic self-consciousness" as the best means of gauging the diverse or disparate ways of modern man, and of combating the disintegrative tendencies; for it sees all attitudes in perspective, it is charitable without being gullible, it provides a vocabulary of persuasion and coöperation. He writes:

> The progress of human enlightenment can go no further than in picturing people not as *vicious,* but as *mistaken.* When you add that people are *necessarily* mistaken, that *all* people are exposed to situa-

tions in which they must act as fools, that *every* insight contains its own special kind of blindness, you complete the comic circle, returning again to the lesson of humility that underlies great tragedy.

You return, in other words, to the Primal Curse; but it is in fact the lesson of almost all the great comedy and tragedy in world literature.

3. REDISCOUNTS

Now the immediate objection to completing a circle is that it leaves one where he was at the outset, wiser but perhaps also wearier. There is the question, too, of just where on the circle he is. Readers are likely to get a little dizzy following Burke himself; they may reach the end of his books with a great fund of stimulating ideas but without a very clear idea of just where *he* is, and how he got there. "You may legitimately complain," he wrote of William Empson, "that often his perceptions are too refined, leading him into a welter of observations that suffer from lack of selectivity and drive. He is far better at marginalia than at sustained exposition." This is an exact statement of Burke's own limitations. It is also a statement of the dangers and difficulties of any effort to embrace, define, and order the many perspectives in contemporary thought. And so the disinterested critic should first pause to gauge his own position, to consider the consequences of a disinterestedness that is nevertheless a form of interest.

His immediate embarrassment is the necessity of constantly announcing the obvious. He may feel a little silly, like one conversing with a deaf person; the repetition and the raised voice make him more conscious of the banality of his remarks. Likewise he runs into stylistic difficulties. Strictly, he cannot come to the definitive period, the full stop; he must constantly resort to the explanatory or apologetic parenthesis, employ such depressing locutions as "on the other hand." He cannot often let go, or in ringing terms call upon men to beware of ringing terms. And though he may complain that there are too many styles like Macaulay's, in which no one can tell the truth, he has to admit not only their satisfactions but their practical value in advancing important ideas and attacking reprehensible ones. He has himself to face this same problem of timing or stress. In an everlasting welter of half-truths, there is at any given moment need of emphasizing what is true in some, what is left out in others. Utter fairness is impossible if only because it demands a measure of strict unfairness.*

The scrupulous humanist has therefore to walk a tight rope, before a noisy audience that prefers the flying trapeze, and while performing this feat he is assailed by all the familiar charges: that he is a demoralized opportunist, a bankrupt idealist, an intellectual sissy—ultimately that most miserable of wretches, a liberal. Moreover, he needs to take the charges seriously. The mere fence-sitter looks as if he were balanced and taking in a large prospect, the navel-viewer may have an air of serene wisdom. In conscientiously looking at all sides of a subject, the humanist may never come to grips with it. He may forget that suspended judgment is itself a judgment, a policy which like any other has consequences, or that a policy of intelligent shifting can slide into mere shiftiness. He may have too great a fear of dogma or shrink from the least sacrifice of interests and values, whereas any effective social organization—or even philosophical synthesis—demands some sacrifice. He may become so habitually restrained as to justify the doubt that he is restraining anything in particular.

At best, the pragmatic test of humanism is extremely difficult to apply. A conviction of the absolute truth of a religious, moral, or political idea unquestionably heightens its effectiveness; as unquestionably, these absolute truths have bred not only lofty idealism but obscurantism, bigotry, violence, tyranny; and how can one be certain, even of religion, whether the total effect has been for better or worse? Furthermore, the stress upon what "works," ideas as "weapons," may lead to a cheap expediency, even to the brutal cynicism of *Realpolitik*. If the humanist is not liable to this latter extreme, he is liable to a neglect of the objective validity of ideas, the stubborn, inescapable *given* in experience that makes a scientific truth absolutely truer than a belief in witches and fairies. Although

* To illustrate from my own experience: in my book on the modern novel, written some years ago, I noted the obvious limitations and excesses of D. H. Lawrence's work but ended by stressing its values. I felt, rightly or wrongly, that in a science-governed age both literature and philosophy needed his impassioned rendering of the "unknown modes of being" and his exaltation of old ways of feeling. Today I should stick by almost any given sentence in the chapter. Yet I should also shift my emphasis and dwell more on the dangers of Lawrence's attitude. In Hitler's world there are men enough to glorify the Unconscious, rally behind the instinctive and irrational.

James, for instance, was a man of principle, he was a little vague about the objective criteria for principles, a little too hospitable to wishful preferences in determining what faiths or beliefs worked. "There are *some* instinctive reactions," he wrote in *The Will to Believe,* "which I, for one, will not tamper with"; it is not plain why he drew his lines where he did. And Kenneth Burke seems a bit too dexterous and fancy-free as he juggles the various possible frames of reference. In his very subtle management of the exigencies of his argument, he tends to slight the exigencies of fact, the discipline of scientific inquiry, and does not always provide solid enough ground for his final preferences.

Yet such admissions also imply that the humanistic, pragmatic, or "comic" approach need not end in an aimless eclecticism or an irresponsible opportunism. Really to harmonize and integrate, the critic must have a direction as well as a basis of operations. He must assert ruling principles, he must use them, he must assume both their logical and their moral responsibilities. Moreover, to conceive principles as tools or hypotheses rather than fixed ends or absolute laws need not weaken their effectiveness. A tool is something useful, and likely to be used carefully; men naturally keep the tools of their craft in good repair. Hypotheses are likely to receive much closer attention than fixed truths, which can take care of themselves, and they have proved efficient enough in science. As for casuistry, a deal of it is inevitable; to state an aim is to acknowledge that the cards are stacked. In any event the humanist has to take a stand in an untidy, unfinished world, and to defend it he will naturally "make the best of it." But the dogmatist is even more vulnerable to such charges. He has to defend his absolute truth at all costs, whatever the conditions and means; he therefore comes to disregard the actual conditions and to justify any means. It is the man of fixed principles, as the recent history of American Communism shows, who is most apt to become profoundly unprincipled.

At the same time, the admission of relativity and uncertainty does not land us in solipsism or spiritual chaos. The common fears on this score are specters conjured by old habits, and they correspond with neither our actual knowledge of the world nor our actual experience in it. Science begins and ends with the admission of ultimate uncertainty—and it remains our surest means of knowledge and effective control. In everyday life, though men like to play the sure thing they also like to gamble —and like it or not, they have to. Even the absolutist lives in a world of contingency and does not pretend to complete assurance in all his decisions. But men do not have to gamble blindly. They can draw upon common sense, funded experience, tested knowledge. The trained critic, it would seem, ought to be able to manage at least as well. Similarly the admission of relativity and multivalence does not make value illusory or judgment futile. It shifts the issues to firmer grounds, in fact removes the issues from the realm of illusion—the realm of eternal truths that turn out to be so strangely contradictory and so temporal—by bringing them within the scope of intelligent inquiry into actual conditions. Meanwhile some judgments are plainly better than others. The assumption of the humanist is that those are likely to be best which embrace the widest range of knowledge and experience, which permit the freest play to possibilities of value, and which are informed by a spirit neither dogmatic nor merely skeptical but always critical.

In short, this attitude is not a rationalization of a timid or genteel incapacity for conviction. If it is easy to suspend judgment, it is even easier to join a parade, climb on a band wagon. The fiercer Marxists might be said to have a very bourgeois hankering for security and certainty; their "iron laws" are as soothing as the softest Victorian sentiment, for they guarantee that the great modern drama will have a wholesome moral and a happy ending. In general, few dogmatists have thoroughly *earned* their unqualified convictions, many are tender minds in need of such a protective shell. But a thorough-going humanism is an arduous discipline. If it can give an exhilarating sense of freedom, endless possibilities of adventure and laughter and dream, it also demands both humility and toughness of spirit: a willingness to question one's own gods and risk one's soul, to admit that one's most cherished sentiments and values have no absolute sanction or insurance, to live with doubt instead of blessed certainties—to live the hazardous life of a mortal in a world neither brave nor new. Thomas Mann is the striking symbol today. He has achieved the broadest, most harmonious synthesis of living writers, but only after a drastic cross-examination of his deepest pieties. No other has more thoroughly, painfully earned his thought and feeling.

Through Mann, finally, I come to the most difficult of the issues raised by this attitude. His failure, for some years, to declare himself definitely in the struggle against the Nazis laid him open to the usual charge of liberal anemia, even to hints of cowardice. He is himself now in the clear, having become not only forthright but militant; yet the general issue remains. Just what is the critic's obligation to the present? At what point, as a responsible citizen, should he surrender his detachment, enlist in the service of some immediate cause, commit himself to the simplicities necessary for effective action, resign himself to the injustices involved in all action? At what point, simply in the interest of his effectiveness as a thinker, should he be positive, unequivocal, indifferent to all the qualifications that fastidious reasonableness demands?

I should not attempt to locate the point. One can trust the brute pressure of events to obliterate undue niceties; one can scarcely hope to maintain a consistently humanistic attitude during a great war. At any time the habit of discount and doubt is clearly not for the idealist who would preserve his single-mindedness, nor for the artist who would preserve his innocence; the high priests and the singers of causes need to see their cause whole and holy. Neither can this habit be the entire policy of practical moralists or of administrators, who have to cope somehow, here and now, with a vast deal of ugly, recalcitrant fact. They have to forego the luxury of the long run for the strain of the short haul; the appalling capacity of collective man for stupid, blind, self-destructive behavior is not to be confronted merely with a recommendation of catholicity and sweet reasonableness, and not to be written off on a check dated in the remote future. Altogether, the humanist must recognize the normality, the practical necessity of the very rigors he is trying to soften and correct. He can hope to mediate and to shape the ends of his society; the partisan supplies the driving force and does the dirty work.

In other words, the final task of the humanist is to rediscount his own discounts. "The march of Montaigne," writes Fernandez, "is a retreat in good order, firm and prudent," to the point where human dignity can be defended and no further concessions need be made. It is a wise and even a noble attitude, it exemplifies "the heroism of modesty"; but it is still a retreat. There is also the heroism of the reckless dare, the arrogant pretension, the impossible idealism. Montaigne himself acknowledged the higher pragmatism that justifies the foolhardy refusal to retreat: "Events are weak testimonies of our value and our capacity." Most great achievement, the very vitality of man, springs from immoderation and immodesty. And in a mad world, perfect reasonableness may be an unreasonable attitude.

Yet men still have to think, even in a mad world; and I should still insist that the crude metaphors, the cartoon categories, the billboard slogans, and all the violent simplicities of the crusader or pamphleteer are inappropriate for the social philosopher or the literary critic. In the realm of thought, an uncompromising devotion to oversimplified means and ends is profoundly unrealistic. Men can always be trusted to be dogmatic enough for all practical purposes; today we have plenty of zealots on our hands. There is the more need of free, disinterested, catholic criticism: the effort to inquire into first principles and final issues; to formulate, with full self-possession, the ends and means without which action becomes still more unruly and unjust; to preserve the values that alone can justify conflict.

> In a day [writes Lionel Trilling] when intellectual men are often called upon to question their intellect and to believe that thought is inferior to action and opposed to it, that blind partisanship is fidelity to an idea, Arnold has still a word to say—not against the taking of sides but against the belief that taking a side settles things or requires the suspension of reason.

Hence all this is not to relieve the critic of social responsibility. It is to define his kind of responsibility. Criticism and creation, like theory and practice, are also to be integrated, and may be ideally united in the person of a genius—a Goethe. But simply to integrate them we must remember that they are not identical. The requirements for a sound social philosopher or scientist are not the requirements for an effective political leader, the attitudes proper for a critic are not always the attitudes proper for a poet. As Kenneth Burke concludes, "Whatever poetry may be, criticism had best be comic."

II

ASSUMPTIONS AND KEY TERMS

It is of great use to the sailor to know the length of his line, though he cannot with it fathom all the depths of the ocean; it is well he knows that it is long enough to reach the bottom at such places as are necessary to direct his voyage, and caution him against running upon shoals that may ruin him. Our business here is not to know all things, but those which concern our conduct. JOHN LOCKE.

1. A NOTE ON DEFINITIONS

"THE one great Bible which cannot lie," wrote Froude, "is the history of the human race." Unhappily, there remains the problem of exegesis. The text is cited to support all sides of all questions; history teaches everything, history teaches nothing—these are head and tail of the same idea. The first impulse may be to cry tails. The record of the best that has been thought and said is also a record of endless contradiction and confusion. Yet one may get a simple lesson from it, for most of the confusion may be tracked to a single source. This is the arbitrary assumption that is taken for self-evident truth. It is an assumption involving other unconscious assumptions which, when explicitly stated, prove to be highly questionable or even demonstrably false. It is the universal habit that makes the whole process of proof seem incidental; for the argument implicitly assumes the idea to be proved, the conclusion is given at the outset. The chief cause of mischief throughout the history of thought has not been aces up the sleeve but jokers in the hole.

All human thought and behavior are necessarily based on certain assumptions, and the most fundamental of them—such as the existence of an external world—are not in themselves strictly logical or illogical, capable of absolute proof or disproof. Common sense is itself a metaphysic, learned before kindergarten, and a partly outmoded one at that; its "plain truths" are premises taken from Aristotle, Euclid, Augustine, Newton, Descartes, Kant, and other sources that might distress as well as surprise the "practical" man. Science, too, begins with a leap into metaphysics. But the recent revolution in science has resulted from the exposure and criticism of its basic assumptions—the notions of absolute time and space, for example, that Kant regarded as necessary modes of thought—and from the recognition that they are strictly *assumptions.* In all thought, more important than the premises themselves is this clear awareness of them and behavioristic attitude toward them. What goes without saying is what especially needs to be said. And so a thinker's first effort should be to discover and drag out into the open all his silent premises. He should lay his cards on the table face up, so that he as well as his readers may look at them.

Hence we approach the problem of definition, the terms in which the assumptions are buried. We can seldom be sure that we know just what a man means by his key terms, or that he himself knows; obviously we need to know if we are really to get together. But it is not enough simply to define our terms. Definition is so important that we also need to have a clear idea of just what *it* is, its own nature and purpose. As Irving Babbitt said, we need "to define the limits of definition itself."

To Aristotle, definition was not merely a verbal process or a useful tool of thought; it was the essence of knowledge. It was the cognitive grasp of the eternal essences of Nature, a fixed, necessary form of knowing because an expression of the fixed, necessary forms of Being. These eternal essences and species have turned out to be pretty perishable, or at least have been brought into disrepute by evolutionary theory. Aristotle's logic, however, is still lusty. His forms of knowing have outlived the Being they were designed to express. For men continue to talk and think in terms of essences. They seek definitions—as of poetry—on the assumption that a single or absolute quality is involved.* They give or report meanings as if they were telling

* A. Ingraham gives a good example in his *Swain School Lectures* (quoted by Michael Roberts in *Critique of Poetry*): "We do not often have occasion to speak, as of an indivisible whole, of the group of phenomena involved or connected in the transit of a negro over a rail-fence with a melon under his arm while the moon is just passing behind a cloud. But if this collocation of phenomena were of frequent occurrence, and if we did have occasion to speak of it often, and if its happenings were likely to affect the money market, we would have some name, as 'wousin,' to denote it by. People would in time be disputing whether the existence of a 'wousin' involved necessarily a rail-fence, and whether the term could be applied when a white man was similarly related to a stone wall."

what something really and truly *is*. "Religion," somebody has said, "is a device for keeping people quiet while skinning them." Somebody else has said that it is "the sum total of man's impulses toward good," and still another that it is "the sum of the scruples that interfere with the free exercise of human faculties." They would all be contributing something to understanding—except that they usually mean "Religion *is* so-and-so," not "I am defining religion as so-and-so." When a mathematician declares that *x* equals something, he is consciously assigning an arbitrary value, and can proceed to carry out useful operations. When a critic defines poetry or romanticism or naturalism, he usually assigns an absolute value, believes that he has uttered an eternal truth, and then begins to fight with other critics. Similarly men still have the primitive faith in the magical efficacy of the right name: once you give something a name you have it by the throat. The man on the street believes that he has explained behavior when he labels it "inferiority complex." The sophisticate coins fancier paraphrases that smack even more of mumbo jumbo. The intellectual above all is apt to manipulate verbal symbols in the belief that he is dealing directly with facts.

Now words may heighten or amplify consciousness in subtle ways, apart from their official value in the currency of communication or even their sound effects. To specify and name a quality of experience is often to *feel* it, to possess more fully as well as to see more clearly. Especially with the youthful mind, vocabulary building is an extension of consciousness; once a youngster has really got hold of such words as vulgarity, sentimentality, magnanimity, his perceptions are apt to be finer and his judgments sounder. Any clear definition, moreover, is useful. Among the most useful, in fact, are the purely verbal inventions that have no objective referents. The head-in-the-sand ostrich, Stefansson remarks—the ostrich-by-definition, comparable to the mathematician's triangle—is in many ways more serviceable than the real bird, and in much less danger of extinction. In the last analysis, all scientific definitions are of this kind, involving some imaginary or strictly unreal terms. But the scientist knows that they are arbitrary. He does not mistake the hypothetical for the real bird.

In short, definitions are properly means of knowing, not ends of knowledge. If they are the end products of a given inquiry, they are nevertheless simply convenient formulations, the means to further inquiry; they do not solve a problem but set it. The literary critic above all must remember that they are merely instrumental. His business is to get at the individual work of art, which is a unique combination of concrete particulars; for him the particular is indeed the thing. Hence he must remember as well that no definition can fully comprehend the real bird. Hence he must be on guard against the severely formal definitions that are so efficient in science; if criticism has notoriously suffered from loose talk, it has suffered even more from narrowness and rigidity. Altogether, the critic's terms can never be exact duplicates or complete inventories of the experience he deals with, and they serve him best as they simply lead him to this experience, give general directions for dealing with it. He wants suppleness and sensitivity, ease and sureness in finding his way around. He therefore does well to travel light, with a few guiding principles but without a complete set of fixed categories.

The last word about the whole process of definition is that there can be no last word. Words generate words, definitions inevitably involve other undefined terms, and thought ends where it begins, in the undemonstrable, the unthinkable, the literally unspeakable. Modern scientists are consciously setting out from the assumption that their basic terms—"order," "structure," "relation"—cannot be defined; or what amounts to the same thing, that the indefinable can *only* be defined. Alfred Korzybski therefore maintains that the first demand upon a thinker is not "Define your terms" but "State your undefined terms"—which is another way of saying, "State your assumptions." Here, at any rate, I propose to make only a general statement of my main assumptions, give only a broad definition of my main terms, so as to mark out my field of discourse but not build a wall around it. And though I shall naturally attempt to make as cool, precise references as possible, I do not hope to avoid all the traps of words. The critic is bound to fall into some, given all the variables and immeasurables he has to deal with, and his readers, with their different sets of associations, will fall into others despite his best efforts. With the old words in particular, possession is nine points of the law; redefine them as carefully as he will, the old meanings will creep back. But he may at least hope to remain conscious of the limits and dangers of language; and this consciousness, I believe, need not become a paralyzing self-consciousness.

2. A NOTE ON VALUES

MOST men, I suppose, would accept the broad statement that the final concern of literary criticism is values, and that a value is the satisfaction of an interest or desire. Most would also accept the assumption that values are grounded in human experience on this earth. Yet the assumptions implied in their thinking are usually quite different. In justifying their choices, men habitually seek some kind of capitalized authority for them, appeal to a disembodied something known as Pure Reason, talk as if they filled their stomachs or fell in love for reasons of heavenly hygiene. They manage somehow to get themselves transported beyond life. Especially when they become formally philosophical, they enter a world supposedly more real, permanent, distinguished than the world of natural processes or "mere appearances." Thus C. E. M. Joad tells us that values constitute a third order of the "subsistent," the other two being matter and mind; he reserves "reality" for this "realm of changeless and eternal objects, truth, goodness, beauty, and, it may be, deity, which are neither mental nor material."

Now I always find it hard to enter these higher orders of subsistence and to follow this trick of pulling ethereal rabbits out of invisible hats. But even if the mind can contemplate something that is neither mental nor material, there remains a solid objection to such speculation. Invariably it leads to futile argument over what is the "real" or "true" value of something or other. The same thing can evidently satisfy different interests in different men; one prizes a painting because of its formal composition, another because of its cost, another because of its association with his dear old grandmother. I accordingly prefer to say with R. B. Perry that a value is "any object of any interest" and that every value is real—even though it is later disapproved because it conflicts with other interests. One has to make choices, of course, and one may usefully classify and distinguish. John R. Reid, for instance, makes out three kinds of value-experience: the "substantive" or simple feeling, as of well-being; the "adjectival" or feeling-tone, as in the esthetic perception of a quality in the outside world; and the "adverbial" or feeling-attitude, as in the pleasure of striving. It is a helpful distinction if only because many earnest men deny or disapprove all but the third kind, seeing in the immense traffic of life nothing but red and green lights. But it is helpful only when the differences are not made invidious or absolute. In trying to measure and order values, one must always keep in mind that here there is no simple arithmetic or definite hierarchy.

The only positive evidence that something is valuable is that people value it, and there is no value apart from interest or desire, no good apart from satisfaction. All definitions of an abstract or absolute good, independent of life on earth, are inevitably artificial. I assume with Spinoza that we do not desire a thing because it is good, but that it is good because we desire it; I assume that there is no good that is good for nothing. The basic interests or desires may be rationalized and educated, but they are not in themselves rational. Ultimately, nobody can say just why we find certain colors, smells, and noises pleasing; they simply are so. Nobody needs a logical reason in order to eat, play, or make love, or can explain why it is pleasant to do so. "If pragmatism justifies beliefs in terms of purposes," declares Michael Roberts, "it must go farther and justify the purposes themselves." I assume that one cannot go behind the fundamental purposes. They are the only absolute, the given data with which criticism must begin, just as all thinkers must begin with the as unthinkable as undeniable fact of thought itself. An amoeba selects and rejects, fulfilling the urge to live and meeting the demands of its environment. Just so do the values of man spring from the vital impulse and constitute his response to his environment.

My chief objection to such assumptions as Joad's, therefore, is that they obscure the concrete context of goodness. Valuation is a function of an organism that has a definite structure and history, is constantly interacting with a definite environment, lives not merely on but *in* the earth. And the *inter*action is important. In the past, value was usually located in the external object. Philosophers and critics dwelt upon the art work or the handiwork of God as independent of the mortal observer, and so they were led to discover all their eternal entities and absolute beauties. Today, however, it is more often pointed out that values are necessarily subjective. Thus I. A. Richards told us that we should never say, "This object is beautiful," but always, "I have a feeling of beauty inside me." Yet the reference to a "feeling" of beauty, to something "in me," is also an abstraction, and always to talk in this way would be as undesirable as it is difficult. The natural reference to the object is necessary to understand and control the conditions of value; to forbid this reference is to encourage more sloppy adventures of little souls among masterpieces. In any event, an absolute separation is impossible. The object produces the feeling, and is also made by it; the subjective is also objective, because part of the natural world.

Similarly the concept of interaction enables one to escape the ancient dualism of the world of matter and the world of spirit—and the as ancient problem of how to get the twain to meet. To be sure, some kind of dualism is a practical necessity of thought. One cannot really escape the notion of subject and object, oneself and something outside oneself. Memory, perspective, the delayed report of sound, the recognition of the same object by different observers, sense deficiencies and hallucinations, dreams, fancies, errors—all these plain facts, A. O. Lovejoy points out, naturally lead to a belief in the existence of an external world, and then to a dualism of mind and matter. Yet this mode of perception does not necessarily imply an absolute separation or antithesis: mind versus matter, man versus nature. Neither is it necessary to put one of these abstractions inside the other, translating subject-object into subject-predicate: mind *is* matter, matter *is* mind. In formal terms, I assume an interpenetration of "mind" and "matter," a continuous interaction between the knower and the known, a sensational mass of which subject and object are different aspects. For practical purposes, I am content with my direct experience, which suggests that there is something solid out there, that I am not a ghost, and that something goes on between us.

How we can know the external world remains a mystery. That we have knowledge of it is a fact, and no fact is better established; the mystery of thought and knowledge is more natural than the logical efforts to explain it—and less mysterious than most explanations given by philosophers. As for the *Ding an sich,* I see no necessity of positively declaring oneself. I waive all questions of the essential what and ultimate why, as

I waive all question of whether human purposes imply some larger or holier purposes. It is salutary to wonder, but it is also wise not to worry about nature's private life. I am content to rest in the facts, conceived as the immediate data: what is directly known aside from the question of what it represents. These facts are not "raw"; our concepts modify even our direct perceptions, also give us "the given." But this further complication leads back to my basic assumption, that human knowledge is a human interpretation. One may describe the motion of the stars in mathematical terms, or one may arrange them in pretty constellations; both expressions serve human purposes and both have only a relative validity. Hence I use the word "truth" always in this pragmatic sense. Truth is disciplined by the external something called reality; the practical success of our dealings with the world makes it impossible to deny some correspondence between this truth and this reality; but to what degree they correspond no one can be sure. In any event, the truth known by man is also created by man, in his own image and for his own sake.

This is an anthropocentric attitude, of course, and may therefore appear primitive, naïve. But as a purposeful creature man is inevitably anthropocentric. Although in his effort to understand the world he should properly try to exclude his hopes and desires, the effort itself is rooted in them. The search for truth is a practical activity, dictated by human needs and guided by human standards. Intellectuals who exalt truth above life itself belong with the worshippers of the crocodile. Man seeks it in order to control his world, to enjoy the pleasure of contemplating it, or in some way to bend it to his living purposes. In a word, whatever he believes, he actually behaves by anthropocentric standards. Theoretically he can get outside himself and become a citizen of another world; actually he is always with himself and busily exploiting this world.

In short, my premises are broadly empirical, pragmatic, naturalistic. "That which promotes the fulness and adequacy of life is that which is valuable," writes H. S. Jennings, the biologist; "and there is no other basis for the concept of value aside from its reference to life, no other basis for the notions of right and wrong." This is still an assumption, not a self-evident truth. But it has the immediate advantage of being more clearly *forced* by the outside world in practice, less arbitrary than any supernatural assumption. It also provides the most solid basis for evaluation. The trouble with absolute or transcendental good, as known intuitively or deduced from some a priori principle, is that it is usually assigned a spiritual domain of its own, quite apart from the crude affairs of nature and practical life. Its devotees encourage what John Dewey calls "the impossible attempt to live in two unrelated worlds"; they cling to "an unreal privacy of an unreal self." Holily they aspire to schizophrenia. And their endeavor is the more unfortunate because their spiritual values can be accommodated in the natural world. As Santayana says, "Naturalism may . . . find room for every sort of psychology, poetry, logic, and theology if only they are content with their natural places." One object of this study is to locate these places.

3. THE NATURALISTIC BASIS OF VALUES

WE begin, then, with a world of living organisms. The primary object of them all, from the amoeba to man, is adaptation to their environment and mastery of its conditions; in the long evolutionary view, their main drive is to proliferate and multiply, to live more abundantly and more complexly. No one knows exactly when, how, or why the device of consciousness evolved in the course of this drive, or where its development will end; but its effect is plainly to enable more sensitive, flexible, resourceful adaptation. In this view man is the "highest" animal because he represents the farthest advance as yet made in the drive of all life. He is by all odds the most complex and plastic organism, he has a far wider range of adaptive behavior, he can exercise far more control over his environment. He alone has learned the habit of learning, he alone can reckon the power of reckoning. Other animals deal satisfactorily enough with immediate conditions—as their mere existence shows. The anthill is indeed a marvelously intricate, efficient organization and might stir a wistful envy in Henry Ford or Stalin. But it is also stereotyped, limited, static. Man is immeasurably more resourceful and flexible, he can extend his possibilities, he can *create* his environment.

Here is the biological basis for efforts to define and establish the good life. "Ripen! Ripen!" cried Sainte-Beuve. "As a man grows older he rots in some places and hardens in others, but he does not ripen." But ripeness means a complete realization of potentialities, and the distinctive potentialities of man lie in his nervous system, more particularly in the cerebral cortex. Hence that is valuable, generally, which heightens, extends, and refines consciousness, and thereby increases the significance of experience; the traditional humanistic ideal of maximum consciousness, a full, harmonious development of human faculties, is a moral expression of the biological fact of growth. Hence that is bad which cramps, blunts, distorts—which prematurely mechanizes adjustments to environment and limits the possibilities of experience. In this view one can make out clearly the source of value in science, art, religion, philosophy, and all the interests and activities we call civilized. One can also make out the source of their abuse: the excesses that make for narrowness, disharmony, incompleteness. And one can understand the irreversible processes in culture and nature: why men who have learned to appreciate Shakespeare and Beethoven can never go back to Joyce Kilmer and Nevin, just as the modern scientist can never go back to the physics of Aristotle.

"We can get from life only what we put into it." No doubt Edgar Guest has made this remark. But these are Goethe's words, spoken to Schopenhauer, and like most truisms they contain a neglected truth. The sensible course would then seem to be greediness: to put in as much as we can get away with. And however misguided their efforts, this is what men naturally try to do. They naturally aspire to abundance of life, fullness of being. If they are also inhospitable to new possibilities, inclined to regard the good thinker like the good child as one who causes the least trouble, they have in the end clung to every extension and refinement of consciousness. Although philosophers, as William James remarked, have always considered the unity of things more illustrious than their variety, what the intellect actually aims at is neither unity nor variety for itself but *totality*. For the universe is less snug than men would have it, often recalcitrant to their purposes; it is also as exhilarating as they make it, wonderfully adaptive to their purposes. It encourages the effort to put a lot into life. It handsomely rewards the assumption that the sky is indeed the limit.

Now men are reluctant to give up such fine names as Soul. Over the centuries they have grown so accustomed to this ethereal raiment that they feel naked without it—even though rude thinkers have told them all the while that they had nothing on. Nevertheless body and mind cannot be separated empirically, and they can get along by themselves. For the concept of soul does not "explain" our behavior; it merely baptizes. Worse, it confuses or obstructs physiological explanation, it discourages the effort to live as biological whole men. Over the centuries thinkers have not only separated but opposed man and nature, erecting marvelous edifices of thought on an elementary contradiction in terms. They have restricted nature to "gross" matter that in its subtlest form is still mere "flesh," the source of all vile impulse. They have attributed all noble impulse to "spirit," an unearthly something that works to suppress the messages which somehow get through from the flesh. They have talked as if their ideals, the very standards by which they condemn nature, were not also natural products, the fruits of experience on this earth. Even many who have given up the supernatural still have a horror of the natural—hanging on to the Devil as they give up the Ghost. "Man becomes good in reality not by obeying but by resisting 'nature,' " Irving Babbitt declared. By "nature" he no doubt meant the beast in man; but he never made clear where the resistance comes from.

We naturally want elevation in our ideals. But we can get it naturally. Nothing in our knowledge of the nervous system is incompatible with our idealism; much of our knowledge suggests even more flattering ideas than we get from most authorities on spirit, who are wont to remind us constantly of its gross landlord, flesh. Physiologists tell us that potential in the structure of the ordinary man's brain is a wider range and finer discrimination of behavior than even genius has achieved, possibilities of experience not dreamed of in most philosophies. At the same time, they do not invite us to leap on impulse and ride furiously off in every direction. Desires are not born free and equal, the full life is not a perpetual spiritual holiday. Moralists who are frightened by this biological language, as a dangerous concession to the animal in man, can be assured by any biologist that the key to efficient behavior for all living organisms is control, that the privilege of being a "higher" animal is also a greater responsibility, and that the privilege of being civilized involves still more duties and penalties. To the child of nature as to the child of God, freedom can issue only from a clear perception of necessity, a steady hold on impulse.

Hence philosophical criticism, the evaluation of values, is a biologically natural outgrowth, as an effort to determine which satisfactions are deepest and richest, most lasting and secure. If naturalism is often made the excuse for moral rackets, no philosophy is safe from the racketeer or foolproof against the ardent disciple.

The nervous system appears to be the source of consciousness, and consciousness is the seat of all significance and all value in human life; yet there is another important distinction to make. Man, writes Paul Valéry, has extracted everything that makes him a man from the defects of his organism: "All emotion, all sentiment is a sign of defective construction and adaptation." This is Pure Reason speaking, a little supercilious as usual about the vulgar company it is forced to keep. Again, however, we are not told where it comes from, or where it gets the authority to pronounce everything else defective; and the naturalist must insist that reason is not of immaculate conception. "Its great original sin," Santayana observes, "is its denial of its own basis and its refusal to occupy its due place in the world, an ignorant fear of being invalidated by history and dishonoured, as it were, if its ancestry is hinted at." Reason is naïve when it makes itself the model of the whole universe; the lofty philosophers who identify the Rational and the Real are akin to neurotics and savages, who also confuse the order of their thoughts with the order of the world. It is naïve as well when it pretends to an utter impersonality, above all desire and emotion, for the very claim gives away the heartfelt desire. Such efforts to guarantee our values are honorable in motive. But they are often deplorable in consequence, as they are apt to lead to unreasonable behavior.

My use of the word "reason" is qualified accordingly. A better word for my purposes would be "intelligence"—except that it has already been brought into disrepute by psychologists with a rage to measure something before they have defined what it is they are measuring. ("Intelligence," Thorndike once remarked, "is the thing that psychologists test when they test intelligence.") What is needed, under any name, is a view of the biological whole man, a view in which we can make out the full value of the rational but also the necessity of the nonrational—feeling, emotion, sentiment, desire. The activities of the higher brain centers, known as the exercise of reason, are the most advanced point in man's development, the finest means of adaptation; but they do not by themselves actually run man. They belong to a nervous *system,* which is in turn subordinate to the system of needs and purposes that is the whole organism.

All these general statements are still relatively easy. Trouble begins, of course, when one tries to fill them with specific content. Let us consider every good a real good, and like James try to hang on to as many as we can: let us consider the best whole as "that which prevails at least cost, in which the vanquished goods are least completely annulled." At any given time there remains the difficult decision of what goods to sacrifice and to what extent—a problem further complicated by the added value automatically conferred by the act of sacrifice. There remains the difficult adjustment between wisdom and intelligence: the sober wisdom that recognizes what now is, the lively intelligence that fashions what is to be. Yet the main objective, I think, is clear enough to make this theory of value usable. The naturalistic approach helps one to avoid the many forms of the absolute that are designed to comfort and sustain man but in practice have chiefly confused and distracted him. It helps one to sift the categorical imperatives handed down from the past, separating the practical wisdom from the outworn dogma, translating divine into empirical sanctions. It helps one to welcome the many new possibilities of value opened up by science as by art, to oppose the as many new forms of dogmatism that would constrict or block off these possibilities. It helps to discover and mobilize all available resources of effective idealism.

But it offers no guarantees; and one must face the consequences of making man the judge of all good and evil. Although most thinkers today consider themselves naturalists, broadly speaking, most still tend, at the critical points of their argument, to claim some absolute sanction for their preferences. Even the pessimist may shirk the actual conditions of human life. Schopenhauer's assertion that Will is the basic principle of life, the master of mind, is a reasonable biological hypothesis, if not a simple tautology; his further assertion that the Will is evil and that man must negate it is the ancient trick of making universal laws out of personal prejudices. Evil by what standard? Whence this authority to condemn the basic vital principle? Similarly others impose their own terms upon life and then complain when these terms are not met. They strike the Byronic attitude that Peacock described: "You talk like a Rosicrucian, who will love nothing but a sylph, who does not believe in the existence of a sylph, and who yet quarrels with the whole Universe for not containing a sylph." And the familiar statement that human life is "meaningless" may itself be meaningless for practical purposes. All readers are acquainted with the melancholy picture: of matter as a mad dance of electrons, of life as "a disease which afflicts matter in its old age," of man as a forked form of life that has learned to strut and fret—of the whole witless show playing itself out mechanically before an empty house, the only issue being whether the universe is exploding or running down. But in denying the existence of a consciousness outside the universe, an intelligible purpose behind the whole enterprise, the disenchanted forget that there is nevertheless a consciousness aware of the universe, and that life has a very urgent meaning for those who consciously live it. If man's purposes make little perceptible difference to the universe, they make a great deal to him.

Yet neither is this to say that whatever is is right. We cannot prove that it is better to be a human being dissatisfied than a pig satisfied, even if we alone know both sides of the question. We cannot say that consciousness and thought must be good because nature produced them; nature has produced, and destroyed, all kinds of oddities. Life itself may be only a brief, fantastic interlude in the drama of the universe; or the race may perish and a new species take over. In any event the point of the whole business is not clear. All we can say is that meanwhile we are conscious participants in a going concern which —alas for the revolutionists—we cannot overthrow, and it seems sensible to try to get the best possible returns. Ultimately, philosophy comes down to animal faith. The simple values most universally accepted—vitality, courage, fortitude, and the like —are instinct of this faith. The humanistic values that have developed with civilization—charity, tolerance, loyalty, integrity—have no absolute sanction and must be derived from this faith. Yet we can say, unequivocally, that if it is worth being human, it is necessary to face the implications of being human. The exercise of intelligence, the development of coöperative habits and communal ideals, the effort toward fuller consciousness and riper culture—all such effort that is usually urged in the name of abstract truth or justice becomes a matter of biological health. And if biology seem a chill thing to live and die for, we nevertheless have our natural pieties, a deep sense of the "tragic dignity of history" and of "humility without humiliation." The ceaseless efforts of great men to establish a better society on this earth are a symbol and an incentive for the faith that the proper study of mankind is, indeed, man.

4. THE NATURE OF ART

One legacy of nineteenth-century thought is the sharp distinction between the esthetic and the practical. The world of poetry has nothing to do with the world of affairs, said Rimbaud; the artist has no more part to play in social life than a monk, said Arthur Symons; and Thomas Mann long suffered from the suspicion that they were right. Similarly the common man assumes that "fine art" is something reserved for museums; the "useful" arts he does not consider really art at all. Behind such attitudes lies the ancient sentiment that to connect art with common life and everyday experience is to debase it. This sentiment may be pious, a token of gratitude for the "heavenly gift of poesy." It may also be precious, a sign of snob values. It is in any event a denial of natural continuities, and like most exalted notions about human interests it finally implies that art is ineffectual, irrelevant, unnecessary. I assume that art grows out of everyday experience in the public world, and that the study of it should begin at home. A clue to its nature and function is the original meaning of the word: "skill in performance," "ingenuity in adapting natural things to man's use." Artistry is "workmanship," its product is a "work" of art, and the *poietes* is intimately related to all "makers." A clue for a naturalistic esthetic is the common man's remark upon a fine piece of work of any kind: "It's a beauty."

In general, I start from the basic assumptions of John Dewey in *Art as Experience*. The esthetic is no strange visitation from without, it does not dwell in deep retreats. Rather, it is "the clarified and intensified development of traits that belong to every normally complete experience." In the slackness, confusion, or routine of daily life, experience is a miscellany, a loose or mechanical succession, one damn thing after another. When we "have an experience," however, something worth mentioning, we have integration, dynamic unity, line and composition—an artistic structure. Thus thought and practical endeavor may have an esthetic quality; they become emotionally satisfying not merely because they attain their end but because the end is a denouement, the consummation of an ordered movement. Art is distinguished from them, Dewey adds, in that "the factors that determine anything which can be called *an* experience are lifted high above the threshold of perception and are made manifest for their own sake." It deals directly with the concrete qualities directly given in perception, and it not only apprehends them more vividly and more finely but realizes them in and for themselves—as men do whenever they are simply enjoying and not transacting experience. But rarely, if ever, is experience entirely esthetic or entirely unesthetic.

Similarly art is akin to all thought and knowledge in that it is a grasp of *relations*. Nothing is intelligible until it has been grasped as form or structure, nothing is significant until it has been related to other things; science and art alike are an organization of experience, alike establish connections hitherto unperceived or unfelt. In this sense beauty is truth and truth is beauty: the end of each is the creation of patterned wholes. And *on earth* this is essentially all we know. Underlying consciousness are the basic patterns and rhythms of human life in a natural environment. The temporal rhythms in poetry and music, the symmetries or spatial rhythms in painting and architecture, have their source in the pulsings and cycles of life itself; and these in turn are among the regular sequences that alone make knowledge possible. In all their doings, from their noblest creations to their games and their doodlings, men instinctively follow rhythms and make patterns. Their dissatisfaction with an unruly game, an unfinished job, an imperfect understanding, an unsettled situation, a disorderly life is as esthetic as it is moral or intellectual; for it is the dissatisfaction with a rhythm broken, a pattern left incomplete.

In this view form is not merely artifice, an added attraction or an unnatural constraint. "Forms" are conventional, and may become artificial, irksome. Form itself is the universal element of art, surviving all the many versions of eternal truth. At the same time, it can never be empty of content. "Significant form" is indeed significant, but of something outside—or inside—the shape; it matters because it contains the materials of experience. Nor can it be sharply or absolutely distinguished from content. The esthetic impulse is an impulse to form, only as matter is ordered does it become esthetic, *how* it is ordered determines *what* it means. The means may not realize the intended meaning; the vital union of form and substance does not imply a complete integration or a perfect harmony. But the means determine the end, are part of the meaning.

In this view, too, it follows that the work of art is never really static, nor the appreciation of it passive. Kant defined beauty as "that which pleases without interesting"; philosophers and critics often distinguish esthetic experience as "contemplative" and its peculiar pleasure as a freedom from desire. This description is appropriate to the final impression of harmony and equilibrium, "all passion spent," and to the freedom from immediate practical interest. Yet it misrepresents our actual experience in art. To experience is etymologically like to experiment—to "try out." Fully to appreciate a work of art is to go through it and grow with it. Even "contemplation" without interest is strictly inconceivable, but the whole affair is an active realization and fulfillment, a rich satisfaction of desire. Art achieves balance, order, harmony, it always comes to a kind of happy ending; but this is literally an achievement, an undergoing and overcoming of stresses, tensions, resistances, conflicts. Sentimental, mechanical, or academic art cannot afford a rich satisfaction because the resistances have been too slight, the happy ending has been reached too easily. Such emphasis upon the dynamics of art may lead to too solemn or strenuous an attitude; although it is clearly demanded by the mightier creations—tragedy and epic, cathedral and symphony—it is keyed a little high for "Hey nonny nonny!" Nevertheless the esthetic object is always an event, a history, and always within the whole history of life on earth. Art is a consummation of the processes of adaptation by which living organisms seek constantly to maintain their integrity and equilibrium amid the stress of constant flux and change.

Now there remain important distinctions—else the critic would be out of a job. If the universe were as utterly one as some thinkers would have it, and everything were everything else, we could not get hold of anything; if it is indeed all one in the end, the end is the finish of all human purposes. But too many distinctions are artificial, invidious, or absolute. Form and substance, intuition and intellect, subjective feeling and objective fact, concrete percept and abstract concept—fictions useful for denoting aspects of endless interaction—are sharply separated and then flatly opposed. I assume that one literally *makes* the necessary distinctions, since they are up to a point arbitrary or merely convenient, and that one may safely make them only in a full, steady awareness of the continuities and uniformities underlying all human behavior. If esthetic experience is experience had for its own sake, with no practical end in view, no human activity can ever be wholly disinterested, impractical, unrelated to the total personality; and this in turn cannot be separated from a society. However the arts begin, they end as a communication of meanings and goods. The problem is to do full justice to at once their immediate content and their ultimate consequence, their peculiar quality and their social function. The chief source of controversy and confusion is a neglect of the one or the other.

In the conspicuously formal arts—music, painting, sculpture—the temptation is to regard purely formal values as self-contained and self-sufficient. Especially in this century, artists and critics have sought to define and maintain "purity." "To appreciate a work of art," wrote Clive Bell, "we need bring with us nothing from life, no knowledge of its ideas and affairs, no familiarity with its emotions"; many artists have shied away from not only realistic representation but any stress on subject matter or ideas. Within limits this attitude is a healthy respect for the medium. But the limits are set by the fact that the medium itself is strictly not pure. Nature gives us not merely lines, shapes, and colors, Dewey points out, but always the lines, shapes, and colors of *things;* they are expressive even in direct perception because they carry with them the qualities of things, they are charged with value because of our everyday experience with things. Abstract or geometrical art is still "natural," as its primitive origins suggest, because it is not actually empty of all natural reference.* In general, the purists talk

* It is significant that Wyndham Lewis is now calling for a back-to-nature movement in painting, maintaining that the substantial visual world is much more interesting than abstract volumes, and predicting that the whole school of abstract art will perish of sheer boredom as painters long again for "a big juicy *trompe l'oeil.*" But the best argument for this school is that it has also

of art as a heightening of consciousness; but there can be no vivid consciousness of nothing in particular for no particular reason. Art makes us feel more alive because it heightens something, against some background, for some purpose. It quickens perceptions and appreciations that are significant in the practice of living.

Literature, at any rate, must be the despair of lovers of purity. Its recalcitrance is the more apparent because modern writers have also aspired to be purists. "The most beautiful works," wrote Flaubert, "are those with the least matter"; he was haunted by a desire to write about nothing at all. But he always crammed his novels with matter. Poets have then tried at least to avoid the expression of ideas, beliefs, logical meanings, any suggestion of a "criticism of life": the Imagists merely projected an image, the Symbolists offered merely "possibilities of meaning," the Dadaists scorned even these possibilities. All such methods, however, can give only a pale illusion of purity. At best they are subtler means of conveying ideas: images mean something, symbols stand for something, Dada is a criticism of all life. At worst they betray an unwillingness to accept the medium, and so an essential *impurity*. For language is the primary means of communication, or even of thought itself, and the currency for all the practical business of the world. The art of letters can never get free of the vulgar uses of letters.

Literature, accordingly, provides the best text for the whole issue of the natural and social function of the arts. Throughout history, critics have stressed primarily its intellectual content, the meanings that could readily be detached, translated, and used for moral, philosophical, or religious purposes. When insisting on the correct forms, they still regarded all formal elements—metrics, metaphor, imagery, symbolism—as mere embellishment of these meanings. D'Alembert boasted that nothing his contemporaries said in poetry was not sensible enough to be said in prose; even Goethe said much the same thing: "I honor both the rhythm and the rhyme, by which poetry first becomes poetry; but what is really deeply and fundamentally effective—what is truly educative and inspiring, is what remains of the poet when he is translated into prose." This attitude naturally led not only to a neglect of the peculiar

---

rediscovered or revivified natural forms, helped to make the visual world still more interesting, contributed to a still bigger eyeful.

poetic experience but to a demand for the "right" ideas. Today the whole tendency persists, in revolutionary as in academic circles. We may be sure that it will always persist, if only because such meanings of literature are the easiest to think and talk about. And although it is a flattering tribute to literature as a social force, it leads to notorious abuse of literature as art. It has corrupted innumerable writers, encouraging them to state instead of render, to let the sympathetic reader supply the concrete experience, to substitute ready-made attitudes for direct perceptions, to pretend to a knowledge or belief that *as artists* they simply did not have.

Recently, however, a reaction has carried all the way to the opposite extreme. Some of the subtlest critics, notably John Crowe Ransom, Cleanth Brooks, and Allen Tate, not only distinguish the social meanings and values of literature; they disparage them as "secondary" or "ulterior," rule them out of literary criticism, bristle at the mere mention of them. Although Ransom admits that most literature is "moralistic," a "literature of ideas," he proposes to withhold the "absolute name of art" from such "impure" work. And so one must wonder how many of the world's accepted masterpieces deserve this exalted name—and what of it if they don't. Similarly, in *Scepticism and Poetry,* D. G. James objects to the general concern with overeffects and aftereffects instead of the immediate poetic experience: "The weight of emphasis must always be on the vividness with which we grasp an imaginative object or situation, and not on the quality or value of the other aspects of our mental condition associated with that apprehension—'emotion' and 'attitudes.'" Although it is hard to dissociate these "aspects," James is entitled to his emphasis; it is the indispensable first step in understanding. Yet "vividness" alone is no key to the significance or value of imaginative apprehensions—trivial or childish imaginings can be all too vivid, and diseased imaginings take the prize. We need also to consider *what* is being apprehended, the whole import and consequence of the experience. "If you aim only at the poetry in poetry," T. S. Eliot has said, "there is no poetry." Simply as we value it, we cannot keep the experience pure.

Bosanquet described literature as the art whose medium was the closest to being no medium at all. It might also be described as the art with the richest medium, the art that can achieve the fullest communication of meanings and goods. In any event we cannot escape the consequences of its medium. We may reject the traditional notion that literature is or should be primarily an elevated expression of ideas and beliefs. Nevertheless ideas and beliefs are inevitably implied in literature and induced by it; without them its concrete representations could have no suggestiveness or significance. The important issue, accordingly, is not whether a literature of ideas deserves the absolute name of art. Rather it is the nature, import, and value of the kind of idea expressed in the purest poem. It is the issue of the "knowledge" or "truth" in literature, which has been fundamental for criticism ever since the Greeks made the distressing discovery that their immortal Homer had told "lies" about the gods.

To this issue I shall return at the end: my whole study is an effort to clarify and relate the meanings in science and literature. Meanwhile I shall indicate only my general position. It lies between the familiar extremes—that art gives the "highest" kind of knowledge, even divine knowledge, and that it gives no real knowledge at all. For neither of these positions is tenable on the assumption of continuity and interaction, the assumption that the esthetic is not a flight from ordinary experience but a fulfillment of it.

One can readily understand why poetry has been regarded as transcendent truth. We are "moved" by it, we are "transported"; we feel that we are seeing into the heart of things. And perhaps we are attuned to the Absolute; no one can say positively that we are not. Yet this exalted feeling can be explained satisfactorily enough on naturalistic grounds. Art is always establishing natural relations, realizing natural rhythms; in its greatest triumphs it can stir a deep sense of the community of man and his history on earth, the intimate relation of every event and every life within an immense whole in time and space. At such moments the world is felt to be indeed a *uni*verse, and the Universe is mighty enough to feel like the Absolute. At any rate, there is no positive way of demonstrating the transcendental truth in art, or of distinguishing it from the homelier truths; and there is again a positive objection to insisting upon it. It raises the problem of how to deal with all the dubious or literally unacceptable ideas expressed in the world's greatest poetry. It tends to elevate poetry into a vacuum, to make it immaterial in the full sense of the word.

This is also the objection, however, to the Know-Nothing school of contemporary critics who would divorce art from truth making. Plainly the writer does have some kind of understanding of life. Call it imaginative apprehension, intuition, or what you will; still it is not utterly different from intellectual grasp or abstract knowledge. A shallow or confused apprehension in poetry, a crude or inconsistent characterization in drama, and all such limited understandings in literature correspond roughly to inadequate theories in science or faulty reasonings in practical affairs. Like propositions, Ransom points out, perceptions are true or false; the writer's report is observable and roughly demonstrable, in a pragmatic sense true. If the subject matter of literature is inseparable from subjective feeling, it is not for that reason illusory or unreal; a feeling is always of some natural event, it is never of nothing. If the representations of literature are in the final analysis "imitations," which Plato decided were the more ruinous to the understanding because imitations of mere appearances, by the same analysis so are the statements of science. And if the understandings in literature are of a philosophically undistinguished kind, because they are of the immediate and the particular, they are for this reason especially pertinent. As dealers in abstraction, professional thinkers are always prone to overlook at once the simplicities and the complexities of actual living: the deep sentiments and tangled loyalties, the interdependencies and incongruities, and all the nonrational or irrational behavior that the unpretentious writer takes for granted.

In effect, literature unmistakably is a criticism of life. It not only intensifies consciousness; it deepens, extends, clarifies, and orders the possibilities of experience. Hence greatness in literature is also a matter of size, scope, depth. Vividness of imaginative apprehension, fidelity to medium, freedom from impurity—by no purely esthetic criterion can we tell why Shakespeare is a greater writer than Herrick, and why an epic provides a more valuable experience than a triolet. The major writers are alike distinguished from the minor writers, who are often superior craftsmen, in that they order more of our possible experience, engage more of our possible personality—come nearer the humanistic goal of maximum consciousness.

At any rate, literature is a force in social life that must be reckoned with, above all in a period of wholesale revaluation.

> The first intimations of wide and large redirections of desire and purpose [writes Dewey] are of necessity imaginative. Art is a mode of prediction not found in charts and statistics, and it insinuates possibilities of human relations not to be found in rule and precept, admonition and administration.

Art may also lag behind the march of events, insinuate distracting or deluding possibilities. In either event it canalizes emotion, crystallizes attitudes, shapes objectives. It helps to form the temper and distemper it reflects.

5. THE FUNCTIONS OF CRITICISM

To the man on the street, criticism means faultfinding, an unwholesome activity; to the young student or romantic lover of literature, critical analysis means tearing a work to pieces, committing an outrage upon Beauty; and many famous writers have been impatient of the whole blundering business. "Who ever got inspiration from an accurate knowledge of the text?" asked Renan. Nevertheless I assume that literary criticism is not a superfluous or extraneous activity. Analysis may reduce it to a matter of taste, and this in turn to an elemental liking, as unarguable as a liking for spinach; in theory, we may enjoy literature without benefit of criticism. In fact, however, even a simple poem is a very complex, sophisticated thing, and criticism is an inevitable phase in the rhythm of our enjoyment of it. The immediate experience is conditioned by previous education and enriched by subsequent reflection. Reflection is also the means of conscious continuity, linking a particular enjoyment with other enjoyments. Criticism is organized reflection, providing more systematic and more extensive associations, making possible a still richer experience. It has manifestly helped to conserve value, by consolidating the reputation of the greater works. It also produces value: simply calling a thing valu*able* gives it an added value.

In this sense criticism may be called "creative." Yet it is a different activity from that which creates the work of art. Critics are naturally pleased to think that they, too, are artists. They are always tempted to agree with Croce, that they realize their true function by reproducing the artist's intuitive impression, in a moment of esthetic activity identical with creation. They no doubt have such moments, and may be better critics if they are also poets. Nevertheless Croce does not make clear why the great artist, with his exceptional powers of expression, should need their work at all; there is a certain conceit in the implication that they convey what he failed to, do his job for him. The serious objection, however, is that these flattering notions only insure the perpetual misunderstanding between artists and critics, a misunderstanding sprung from the tacit assumption of most critics that the aims, attitudes, and experiences of the two parties are identical. Intimate as are their relations, the critic properly uses a different vocabulary, cultivates different states of mind, emphasizes different attitudes. He has to be wary, for instance, of the black magic of words, which it is the very business of poets to exploit; he cannot afford to be either so artful or so artless. Let him be lyrically appreciative and he is still not writing a poem. Let him consider how different our histories of literature would be had they been written only by poets, and only for poets.

Criticism is indeed whatever men have a mind to make it; but its peculiar function must lie in what distinguishes it from the creative activity. Its problem is not how to reproduce the work of art but what to do to and for this work. And about this problem there are many minds. For centuries the literary critic has acted as historian, orator, curator, philosopher, high priest, grammarian, doctor, or simple tourist guide. Today he may be a psychologist or sociologist as well. But at all times he is disposed to be exclusive, to assume that a useful kind of activity should drive all competitors off the market. Joel Spingarn's manifesto on the "new criticism" is typical. "We have done with the race, the time, the environment of a poet's work as an element in Criticism"—and with the topic sentence of each paragraph he exterminated another species of critic (lower case). As a matter of fact, we have done with none of them; all the species are still alive and kicking. Nor should we have done with any of them. On the face of it, all this varied activity is pertinent. All contributes something to a fuller understanding, richer appreciation, and finer appraisal of literature. Short of this very general purpose, there is no one "true" aim of criticism.

Hence we might look first of all to the more unfashionable kinds of activity. It is easy to ridicule, for instance, the academic scholars: their community can look like a washerwoman economy, in which they stop taking in one another's footnotes only to breed more Ph.D.'s. Nevertheless their labor is highly coöperative, and their product is not only immediately available but indispensable; they give us our literary history, the background necessary for understanding and judgment. Similarly the inspirational critics, the specialists in adjective and epithet, can help to keep literature alive. They may give the impression that the measure of art is the purity and intensity of emotion aroused, their raptures are often simply embarrassing. Nevertheless they may kindle a genuine enthusiasm and reverence, and not merely in undergraduates or clubwomen; there is room for their wholehearted appreciation of the arts in this hypercritical, hypertechnical age, when so many literary men are given to nervous rationalization and worried apologetics. And there is room even for the bookkeepers and pigeonholers, the simple Adams who name the animals as they pass in review.

A more difficult issue, however, is again the demand that the critic confine himself to purely esthetic meanings and values. I agree that his specialized job in the division of intellectual labor is the study of literature *as literature.* Critics interested in social origins and uses become concerned chiefly with what literature has in common with other activities, use it chiefly as an illustration of something else. They tend to leave literature far behind: like Matthew Arnold, our ambitious critics keep turning into social philosophers. Yet this is not a deplorable development; nor can any of this activity be ruled out of literary criticism proper. Literature is never merely literature, its meanings are never merely esthetic. A complete understanding and appraisal of it necessarily involve its relation to our whole experience. T. S. Eliot states the gist of the matter: "The 'greatness' of literature cannot be determined solely by literary standards; though we must remember that whether it is literature or not can be determined only by literary standards." I assume that the greatness of literature is a proper concern of criticism, that this concern need not lead to a sacrifice of literary standards, and that without it there can be no adequate standards.

The critic may appreciate, analyze, interpret, philosophize, propagate, exhort—to repeat, he may legitimately do whatever he has a mind to. But it follows that we ought to know just what he *is* doing. He is always prone to fall into a confusion of aims, as between that of setting forth an explanation and that of inducing elevated emotions; his readers are always prone to welcome the confusion as a sign of wholeness or wholesomeness. It also follows—depressing as the thought may be—that all criticism calls for further criticism, to supplement when not to discount. All interest limits because it focuses attention, all perspectives have shadows because they are perspectives, all attitudes are angular because they are frames of thought. The soundest critic still needs to be sounded: he may be admirably balanced but not know what he is balanced on, he may be admirably poised but simply stay poised and never take off. "A man writes as he can," says R. P. Blackmur; "but those who use his writings have the further responsibility of redefining their scope, an operation . . . which alone uses them to the full."

Here is again the argument for philosophical criticism, in the light of a humanistic attitude. One cannot perform this operation unless he is dispassionate, catholic, flexible: able to perceive the bearings of the diverse activities and admit their validity, able to take his own bearings and enlarge his own scope. A man writes as he can, and he may write better for a wholehearted devotion to half-truth. Indeed, the most stimulating, original, brilliant critics are seldom catholic or "sound"; they are apt to forget the obvious, they are provocative because also merely provoking. But they above all require further criticism, to make the best use of their original contribution.

This responsibility leads us, finally, to the source of all difficulty in criticism. The great effort of a sincere man, De Gourmont observed, is to erect his personal impressions into laws. The effort can appear as a naïve arrogance: the critic announces that what he now likes, all men must always like—or even that they must never like anything else. It always disposes to a militaristic attitude: he mobilizes a large standing army of arguments and then, like Samuel Johnson, talks for victory. Nevertheless this effort is indeed the mark of sincerity, the measure of all responsible thought. And the issues it raises are especially difficult in literary criticism, to repeat, because the work of art is a highly individualized creation. Let us put the problem at its worst. Limitation, partiality, prejudice are avowed in the mere recognition of *individuality*. Strictly, moreover, one is never even the same individual twice; the organism never responds in exactly the same way, never responds to exactly the same environment. Literature, accordingly, is a host of unique works created by individual artists at a particular time in a particular culture, and experienced by a host of individual readers at different moments in a changed or different culture.

At first glance, the picture does not make sense. The impressionists have therefore denied the possibility of any objective judgment whatever. Jules Lemaître declared that the critic could never go beyond defining his own impressions at a "given moment"; Anatole France declared that he could only relate the adventures of his soul among masterpieces. Yet this attitude makes still less sense. The fancy-free impressionist is inconsistent from the moment he sallies forth for adventures (usually all dressed up and willing to pose for posterity); he begins by tacitly assuming the value of masterpieces, he subscribes to the general opinion that certain works are masterpieces, he relates his adventures because they bear on the general experience—he depends on the plain fact of *common* sense. His fallacy is a very simple one. He forgets that *all* experience comes at given moments, to scientists as well as to art lovers, but that its meaning is not restricted to those moments; otherwise it could have no meaning. He forgets the basic uniformities that underlie the unique particulars: the uniformities in the structure of man and the natural environment which unite all adventurers in a common adventure in a common world, and without which we could not even grasp particulars, define impressions. Significant criticism is possible for precisely the reason that literature, science, history, society itself is possible.

The immediate way to make sense out of the whole picture, then, is as usual to follow the lead of practical activity. The plain man has no trouble going beyond the momentary impression. He can arrive at general conclusions short of universal laws; he can pass sensible judgments and make sensible choices without benefit of absolute standards; he can recognize a superior steak though he has not defined the Ideal Steak or the essence of Steakness. So may the critic generalize and theorize without categorical principles, discriminate and evaluate without infallible touchstones. Like all activity, criticism is an adventure; its history is the sufficient refutation of any claim to certainty. Nevertheless it does have a connected history, and the history can be read with profit.

Thus the first lesson is that standards are necessary but also necessarily general. The critic should take into account the individuality of every writer and the uniqueness of every work, the rich particularity that is the stuff of all art and the source of its peculiar value; the historic tendency has been to smother artists under blanket statements comparable to "I hate Jews," and to use their work chiefly as an instance of abstract principles. Similarly the critic should at once allow for in theory and discount in practice the inevitable "politics of taste"; for the temperamental preferences of the individual ultimately lead to the different preferences of whole societies and whole ages.* Above all, the critic should admit the possibility of different interpretations of a given work, or even a given line; for just this possibility distinguishes the meanings of art from the meanings of science.

Implicit in Croce's doctrine of creative criticism is the assumption that the work of art is static, self-contained, intact, having an objective meaning that can be literally recovered once and for all. Similarly Matthew Arnold believed that the aim of criticism is "to see the object as in itself it really is," and then to give the "real" estimate of its worth, not the fallacious historical or personal estimate. Most critics have explicitly adopted this assumption, and all, I suppose, at some time write as if they have. They naturally try to discover the whole intention of the artist, the whole meaning of his work. Yet they cannot possibly do so.

* Although there can be no substitute for fine taste, even the most sensitive critic needs to base his judgment on more than his personal experience in literature. It seems to me easy enough to qualify one's actual likings or to admit values that one does not actually experience. I get a deeper pleasure from Joseph Conrad and Thomas Wolfe, for instance, than from novelists whom as a critic I should rank higher; I get very little pleasure from some of the world's masterpieces, though I can see why the world considers them masterpieces; and I miss a great deal in all the great galleries and museums. If I am confessing a sad deficiency in taste, I am still confident that no critic appreciates everything as it ought to be appreciated even by his own standards. This discounting of one's experience is admittedly dangerous; it is apt to lead to insincerity, hollowness, timid subservience to convention, timid refusal to admit the real obsolescence of the timeless and the real value of the timely. But some reserve seems advisable when we find so wise a critic as Ransom worrying Shakespeare's sonnets chiefly because Shakespeare was not Donne, and when so many sensitive contemporaries are supercilious about Milton or merely contemptuous of Shelley, poets valued by generations of as sensitive readers. In their anxiety to get at the roots of everything, furthermore, contemporaries often overlook the superficial, undignified, but constant element of mere fashion in literary tastes. The complex history of literature is also the simple story of now in, now out; its course is governed by obscure forces but also by such obvious tendencies as that of men to get fed up with anything. As we consider the various styles of the past, we are apt to forget that our own attitudes are in part simply stylish.

The text is inseparable from a context, and to read it "literally" is beyond the powers of the most imaginative. Men are always building better than they know, or worse. The artist himself does not know all he intended, much less just what he achieved, still less what it will mean to a later age or a different society. "It is the peculiar beauty of all the higher activities of man," wrote Hebbel, "that they serve purposes that were not in the mind of the acting individual." Multiple meaning is the very sign of imaginative creation, multivalence the very test of value. In short, any critic worth his salt will attempt to get at the "true" meaning of *Hamlet*. No critic, save one of colossal conceit, can believe that he has succeeded. If we could recover exactly what Shakespeare had in mind, we would doubtless have something splendid, superior to the experience of any one reader; but it would still be something less than the whole import and value of *Hamlet,* which has also been created by countless appreciative readers in ways Shakespeare could not have foretold.

The critic should therefore be wary of all the current schemes of interpretation that reduce the rich experience of literature to a particular scale of meanings, force it into a narrow frame of reference, and discredit the values that are left over. No formula can explain or express all the manifest values at once in Shakespeare and Racine, Pope and Whitman, Lamb and Voltaire, Fielding and Dostoyevsky. More important, the critic should reconsider the whole problem of judgment. By long tradition, his chief duty is not merely to exercise judgment but to act as the official judge; he seldom appears unhappier than when he is not sure what verdict to hand down. He is also notoriously apt to be a hanging judge; and here is the obvious mischief. But it goes far beyond all the stupid or brutal verdicts, the injustice done to writers who broke the laws, or even the lasting injustice done to writers—such as Corneille—who were unable to realize their potentialities because they tried to obey the laws. The worst mischief lies in the impoverishment of our actual experience in literature, the "trained incapacity" to appreciate or even to recognize the rich possibilities of meaning and value. Although the dogmatic critic's judgment may falsify his experience, it is usually consistent enough with both his perceptions and his principles; his whole response has been narrowed and blunted from the outset. Hence he demands, specifically, that the writer have the true philosophy, express the right ideas and ideals;

whereas the chief source of value in literature is its power to transcend any one philosophy, to suggest or support other ideas and ideals—to nourish all ideation and idealism. In this aspect, the chief proof of its vitality is that it has survived centuries of criticism.

Yet substitute for "judgment" the broader and gentler term "evaluation," and it is indeed the proper business of criticism. It naturally enters esthetic experience, as it does all perception; it is the natural culmination of enjoyment as well as repulsion. "The desire for judgment," Fernandez points out, "by no means implies a renunciation of the wealth of feeling; it means only that man, after having proved his powers, wants to complete and harmonize them by adapting them more exactly to the world and to himself, and that he accepts all the consequences of this exactitude." The trouble, he adds, is that critics commonly "set out to *feel* only what they have already *comprehended*." But even in the narrower, harsher sense, judgment remains an obligation. If we assume that art is a vital need and a positive force, we must look to the satisfaction of this need, face the consequences of this force. Dewey himself is at this point not pragmatic enough. In attacking the judicial habit, he pays too little attention to the practical consequences of bad art. In its lowest forms—soap operas, true confessions, comic strips, success stories, Hollywood romances—it can fix sentimental attitudes, condition a whole citizenry to immature, unrealistic, crude behavior. A decadent or snob art can have as serious effects upon sophisticates.

All this leaves the critic somewhere in the middle, between the extremes of an irresponsible exercise of sensibility and a rigorous administration of law. My thesis remains that here he will have not only plenty of room to move around in but solid enough ground to take a stand on. As self-consciousness is the beginning of escape from the limitations of self, so can he achieve more real command and control by a full awareness of the personal, cultural, temporal factors that condition his thought; thought that pretends to utter objectivity and utter freedom is to that very extent blindly limited and determined. In general, the common man is not troubled by the familiar fact that an object has a different appearance when viewed from different positions. He does not establish its "real" appearance by abolishing perspectives, getting a God's-eye view. He is content to know why it looks different in different perspectives, to realize that any view is partial, and to change his position if he wants another view. No more should the critic be troubled by the thought that all approaches to literature are perspectives, and every insight is also an oversight. To see and estimate the object as it really is is to see it in its social context, not in itself, and to see it with many eyes in many perspectives, not from nobody's point of view. The critic needs to go through, not around, the multiple possibilities. For these hazards of his calling are also its rich rewards.

6. A NOTE ON METHODS OF ANALYSIS

EMERSON remarked that it is a good thing, now and then, to take a look at the landscape from between one's legs. Although this stunt might seem pointless when things are already topsy-turvy, it can be the more helpful then. One may say that what this chaotic world needs first of all is more *dis*sociation; by breaking up factitious alliances and oppositions, one may get at the deep uniformities. Or what this nightmarish world needs is the strategy of the dream, which appears to multiply and magnify contradictions but actually ignores them. ("Dreams are particularly fond of reducing antitheses to uniformity," Freud wrote, "or representing them as one and the same thing.") Specifically, the situation calls for a technique of analysis that Kenneth Burke names "perspective by incongruity."

In its simplest form, this is merely a violation of the intellectual proprieties by mating words that have moved in different circles—as when Mencken described hygiene as "medicine made corrupt by morality." Such bundlings are the essence of paradox and epigram, and a familiar trick of humorists and satirists. They are also the essence of metaphor. And as a marking of unsuspected connections they lead, ultimately, to the heart of all thought and knowledge. The great revolutionary thinkers are those who most violently wrenched traditional associations; Karl Marx was a philosophical Oscar Wilde, more scandalous because more sober. Hence Burke has deliberately, systematically cultivated "the methodology of the pun." Throughout *Attitudes toward History* he uses the religious vocabulary of motives for describing esthetic and practical activities, the esthetic for religious and practical, the practical for religious and esthetic. By such impious means he piously strives to integrate these vital interests. Perspective by incongruity enables the perception of essential congruity.

The lead here is the parable of the pike. Placed in a tank with some minnows but separated from them by a sheet of glass, the pike bangs its head for some time in an effort to get at them. At length it sensibly gives up the effort. Much less sensibly, it continues to ignore the minnows after the glass is removed; it fails to revaluate the situation. In other words, it becomes a dogmatist. For just so are men's powers of analysis and adaptation stupefied by unconditional, is-nothing-but generalizations. Thinkers demand that we choose naturalism *or* idealism, communism *or* capitalism, revolution *or* reaction. In the name of realism they copy the pike.

This approach admittedly offers an easy way of calling people names, and may become merely the sport of an intellectual playboy. Nevertheless the unconditional behavior of the dogmatist is in a real sense pathological. Freud has made us aware of the dynamic, reflex relation between emotion and symbol or idea. These reflexes are up to a point socially necessary; communal effort demands a certain measure of identification, personification, or downright symbolic fraud. The difficulty, however, lies in locating the point and determining the measure. Given the impossibility of general agreement upon what are the "right" ideas or how much emotion is "normal," their intimacy is plainly a hazard. In any event an unalterable association is the sign of the fanatic, ultimately of the lunatic. And here is a clue to the extreme nervous tension in the modern world. Psychologists have been inducing nervous breakdowns in animals by stepping up the demands upon their powers of adjustment and suddenly reversing signals to which they have become conditioned; under such artificial stress the animal becomes jittery and finally goes to pieces. Similarly men break down in a rapidly changing world that demands constant readjustment. Yet these demands need not be so intolerable a strain. Men can maintain efficient reserves, they are naturally flexible, they are physiologically capable of far more extensive readjustment. Only by permitting themselves to become absolutely conditioned to some expectation do they become unequal to shifting demands. There is occasion enough today for viewing with alarm; nevertheless men themselves create the occasion and cultivate the habit of alarm.

The method of "planned incongruity" can accordingly be a practical, social way of making men at home in the world they perforce have to live in. As Burke says, it is a way of making perspectives *"cheap and easy."* There is some real cheapening of quality, indeed; uncompromising devotion to a creed is also a form of virtue and strength. But the gain in efficiency and practical wisdom more than compensates for the loss. Complexity, relativity, multiplicity, flux—the conditions of modern experience cannot be liquidated by any effort of thought or legislative degree. They call for a lithe, sinuous, athletic type of mind. The thinkers who bulge with muscular dogma are like the strong men of the advertisements, very impressive, but a little monstrous; and at that they are no match for gorillas like Goering.

I do not propose to employ this method as systematically—or self-consciously—as Burke. But it is manifestly pertinent to my main intention, of relating scientific and literary meanings and bringing them both back to the public world. More specifically, it will be a means of becoming oriented to modern science. Science always seems a little strange to the layman because in a sense it reverses the common notion of how we come to understand; it replaces the familiar particulars with increasingly unfamiliar abstractions—it explains the *known* in terms of the *unknown.* Modern physics in particular might be considered the most radical perspective by incongruity in the history of thought, for it has resulted from a deliberate violation of the axiomatic or self-evident, a methodical thinking of the unthinkable. And this brings up the ruling logic of modern science.

Bacon remarked the paradox that the term "ancient," with its connotations of mellow wisdom, is habitually applied to the *young* peoples in the history of the race. Actually we are the ancients, the Greeks were the precocious youngsters. Yet contemporaries are still inclined to obey these youngsters, as Dewey has taken pains to demonstrate. However unconsciously, they cling to the Greek concept of fixed forms or eternal essences, by which variety, particularity, change, contingency—all that is most plainly given in direct experience—are made to seem like "mere" appearances, or accidents to be put up with. With this concept of Being they inherit a worry over ultimate Being—what makes action act, or causation cause, or being be. Above all, they still regard Aristotelian logic as the final arbiter of Truth. They subordinate to it the method of scientific inquiry, certainly the most efficient means to truth that man has yet devised. And the working logic of science is not only significantly different but in some ways inconsistent with the ancient canons.*

Bacon also remarked that Aristotle's logic is not subtle enough to deal with nature. For example, it cannot handle satisfactorily the quantitative relations that are fundamental for science. Its subject-predicate propositions have become more inadequate with the complex, multiple interrelations stressed in the important concept of organism. Its law of the excluded middle—a thing is either A or not-A—has recently come in for especial attack. Although this law is up to a point a necessity of thought, strictly it is refuted by life itself; a living organism, always growing and changing, is at every moment itself and something else, and demands as well a law of the *included* middle: a thing may be *both* A and not-A.† In general, criticism centers on Aristotle's static, absolute, immutable categories. In their practice, at least, scientists observe no such categories. They employ an instrumental logic that controls inquiry but is also controlled by inquiry; they will break any law of thought in order to make a law of nature.

Modern philosophers, accordingly, no longer recognize one

* Since my references to Aristotle will usually be uncomplimentary, I wish to anticipate the charges of cocky modernism or professional irreverence. In a historical study, full justice could be done to his pioneering achievement, his whole important contribution to the life of reason and reasonableness. This, however, is a critical study of the present, its knowledge and needs. Aristotle's thought is a great monument; but the trouble with monuments, as Bacon said, is that they can only be celebrated, they cannot be moved or advanced. His basic principles—which are still religiously taught by disciples in our great universities—have made possible no considerable discovery, can generate no new knowledge. Dewey points out, moreover, that the very realism that distinguishes him from Plato has made him a more positive hindrance. He was primarily a student of the facts of nature, the worldly rather than the otherworldly, and he put the Platonic Ideas or eternal forms of Being back into nature. Here, after two thousand years, Darwin finally exploded them; but meanwhile Aristotle's basic assumptions had got into science, where they were unrecognized even by the seventeenth-century pioneers who believed that they had done with him, and they hampered its advance until Einstein. Altogether, our job is to do for our age what Aristotle so brilliantly did for his. We can hope to succeed only by recognizing that thought is a historical enterprise, not a final solution.

† Zeno's famous sophisms are a case in point. His problem of just how many hairs are needed to make a beard, how many grains of sand to make a heap, is a real problem—*if* we hold that there must be either beard or no beard, heap or no heap. The problem disappears the moment we take our minds off mere words and consider natural facts. Similarly Zeno's demonstration that Achilles cannot overtake a tortoise has bothered philosophers for centuries, even though Achilles always can catch the tortoise, because it is logical by the old canons. What the paradox actually demonstrates is that this logic cannot handle continuous change.

fixed, essential logic but are developing a number of logics, including notably one of relations. The intricacies and refinements of these systems are beyond the scope of my present purposes, not to say my attainments. For the moment a simplified version of the new techniques of thinking will do well enough; and such a one is the "dynamic logic" outlined by Boris B. Bogoslovsky in *The Technique of Controversy*. He states four main principles: (1) A must never be used without not-A. The statement that all men are selfish has no useful meaning—or if taken literally, as by the youthful cynic, is seriously misleading—unless we state that men are also unselfish. (2) The references of a word must be explicit. "Selfishness" has various aspects and implications, and confusion is inevitable until we specify the meaning we have in mind. (3) The assumption of continuity and interaction demands the principle of "bothness." Thinkers have long debated such questions as whether men are products of heredity or of environment; obviously men are products of both, and thought must always take into account the continuous process, the inseparable conjunction. Hence (4) quantitative values must be made explicit. We must state *to what extent* a thing is A and not-A (selfish and unselfish), place it on the scale between these poles. Altogether, the aim of "dynamic logic" is sharper specification among multiple relations, more accurate location on a continuous scale. It is never independent of natural knowledge. It can spot the obvious fallacies resulting from reasoning by analogy; it nevertheless recognizes that given a world in which there are no absolute identities, *all* reasoning is by analogy.

Now this is indeed a simplified version of scientific logic, and as outlined will hardly seem revolutionary. Bogoslovsky makes it seem too easy, moreover, especially in his call for quantitative values. Long ago Hume pointed out the inevitable ambiguity in controversies over the degrees of a quality or circumstance; in the issues of the arts and humanities we cannot hope for utterly precise locations and measurements. Yet merely to recognize and define the problem is a considerable gain. Everything is indeed a matter of degree. For the practical purposes of thought it is not all one, as we say; it is always two. If we mark the poles, include the middle, use a sliding scale, we can at least hope to make controversy more profitable; ideas may approach and not merely collide like billiard balls. The assertion that the Heart sees farther than the Head gets us nowhere, until we specify what kind of thing it sees better, under what circumstances, for what purposes—always remembering that heart and head see in conjunction, and are not engaged in a seeing contest. And elementary as these principles may seem, few thinkers consistently apply them. The logic of most discourse is still based on the sheep-or-goat concept of truth.

Thus critics tend to treat as independent, exclusive principles such poles as romanticism and classicism, freedom and tradition, form and substance, dogma and sensibility, individualism and collectivism, self-expression and communication—head and tail. They snap continuities by the use of "versus," making antitheses of complementary principles that are at work in all artists at all times; they take a stand at one end of a scale as if it were the norm, the absolute A of beauty or truth; they make out degrees only in terms of error or evil. They then erect policies into proverbs and proverbs into categorical laws. In effect they demand that we choose between the principle that one should look before he leaps and the principle that he who hesitates is lost. An efficient logic should help us to tell when to leap and when to hesitate, enable us to keep clear of the leaping school and the hesitating school.

All this still amounts to a restatement of a general attitude rather than the formulation of a rigorous program. More often than not it will appear as a demonstration of platitudes, as that the same thing can always be at once good and bad. But through "planned incongruity" and the principle of ambivalence or polarity, one may set platitudes in a richer context, restore the living truth in truism, at times even startle. One may locate the good in the bad and the bad in the good, and then make for better. Thus regionalism in literature (to return to an illustration I have used elsewhere) appears as a relatively simple adaptation to life, a return to the old meanings of the family, the home, the soil. It may therefore be stigmatized as an evasion of the problem of assimilating the complex material of modern life, a shirking of responsibility, an escape. But it also may be applauded as a return to the grass roots of art and life, a recovery of essential pieties, a means of steadying a confused generation. Reviewers tend to use consistently either the "bad" or the "good" description; they speak kindly even of the feeble exhibits or coldly even of the sturdy ones. The ideal critic would command the whole scale of motives and values in regionalism and place a given work accordingly.

In these terms, at any rate, humanism is an effort to place all doctrine on an appropriate scale, to see it in relation and in degree instead of as isolate truth or vagrant error, to provide a perspective in which dualistic aspects may again be seen as aspects of a whole—the organic whole that is the included middle. The yes and no constantly asserted in daily behavior are naturally translated into right and wrong, good and bad; but we can make choices without becoming Manichaeans.

PART II

DEVELOPMENT:
THE IMPLICATIONS OF MODERN SCIENCE

III

AN INTRODUCTION TO MODERN SCIENCE

Let us admit the case of the conservative: if we once start thinking no one can guarantee where we shall come out, except that many objects, ends, and institutions are doomed. Every thinker puts some portion of an apparently stable world in peril and no one can wholly predict what will emerge in its place. JOHN DEWEY.

1. SCIENCE AND THE TRADITION

ALTHOUGH science is no doubt the Jehovah of the modern world, there is considerable doubt about the glory of its handiwork. Apart from such downright heretics as D. H. Lawrence, there is a growing distrust of science, a growing anxiety to minimize and localize it. A large company will tolerate it only on its best behavior, and only for the hewing and hauling, the provision of the material means of life. They will have none of it in the matters of ideals or ends, which they hand over to religion, philosophy, and art. And they resent science because it has confused these high matters. Howard Dykema Roelofs expresses a common sentiment: "Religion can produce on occasion what science never does, namely, saints. Today we have science and scientists aplenty. We lack saints."

Now the easy answer is that religion has been on the job for some thousands of years, and that its record does not inspire confidence. The saints have rarely had much to do with the administration of society; there is some question whether they compensate for all the obscurantism, the intolerance of doubt and dissent, the justification of barbarous means by holy ends, the bigotry, hatred, and tyranny that have always flourished under religion, and often been fostered by it. But such tit for tat gets us nowhere with the real problem. Science has introduced very grave complications into the drama of civilization. The miracles it has worked become means of broadcasting and amplifying vulgarity, greed, and stupidity; worse, they include high-powered engines for wholesale slaughter; and their sum is the problem of managing an immensely complex industrial society. For better or worse, however, science is plainly here to stay—at least as long as our civilization. As plainly, then, the sensible policy is to make its stay for better. And the first move is to drop all invidious comparisons and take an easy look at it. To resent so remarkable a power because men abuse it is as illogical as to resent language because men speak error and evil.

Men rarely marvel at all their practical knowledge. They take it for granted because it always works; they are more struck by the talisman that occasionally seems to work, the prayer that seems to be granted. This attitude is understandable, given all the contingencies and uncertainties of life, but it is also a hangover from primitive times, when nature held many more mysteries and terrors. The philosopher who answers the extraordinary questions inherits the prestige of the witch doctor; the priest who calls for prayer invokes the *feeling* of security that men especially needed when they had so little knowledge or power of control. But the more particular reason why science is held in relatively low esteem is the persistence of an ancient aristocratic tradition.

John Crowe Ransom refers approvingly to the "secret notion" of most art lovers, that art distinguishes Man from animals in kind, science only in degree of efficiency. This is scarcely a secret; artists have long been free enough in announcing that their activity is higher and holier. They can appeal, moreover, to the "inveterate conceit" remarked by Bacon: "that the dignity of the human mind is lowered by long and frequent intercourse with experiments and particulars, which are the objects of sense, and confined to matter." This conceit is in turn, as Dewey has pointed out, the historic product of class relations. In Greek and Oriental cultures, practical activities were the business of artisans and traders, when not of slaves; the exercise of pure reason, all transactions with the rulers of the universe, were the exclusive privilege of the higher class, and were accordingly made independent of the vulgar operations of doing and making. "Theoretical kinds of knowledge," Aristotle said, are "more of the nature of Wisdom than the productive." Ransom clearly has this patrician attitude; throughout *The World's Body* he refers to science as something that is "merely" useful, that "belongs to the economic impulse and does not free the spirit." And a host of literary men are much more supercilious about the wealth and power that enable them to contemplate their inability to live on bread alone.

This tradition explains the curious fact that an advance in knowledge is generally viewed as a "problem." Offhand, one would expect men to welcome science as a means of securing their values; it seems a threat because the official custodians of values have kept them in an unearthly realm of their own. In general, the source of all confusion and distrust is that science, the author of our world, is still a stranger here itself. Its findings have yet to become really absorbed in the tissue of common knowledge, its methods really naturalized in everyday habits of thought. Simple men play gladly with its by-products, but they have not caught on to the changes in the rules of the whole game; intellectuals still approach it like youngsters learning the facts of life, and like the bright youngsters often need to know a little more or very much less. Because old ways of looking at things seem more natural and necessary, and old habits like essential ingredients of human nature, new ways and new conditions of life are regarded as outrages upon human nature, defiances of its constitutional rights; and all present disharmonies look like permanent incompatibilities. Hence science is considered a fearful hazard, if not a blight. Its method seems somehow more "inhuman" than the rationalism of philosophers and theologians; its electrons and fields of force seem more remote from living reality than abstractions like "substance" and "essence"; its theories seem more dangerous than the absurdities men have lived by for centuries; its offspring, the Machine, is simply a monster.* And so men try to solve the problems science has raised, not by a thorough assimilation of it, but by an anxious restriction of it to the "material" or "economic" realm—a kind of ghetto where it cannot contaminate the pure Aryan blood of their values and ideals.

* *The Grapes of Wrath* is a striking example of this neopathetic fallacy. A leading villain in Steinbeck's novel is the Tractor, ruthlessly ploughing up the lands and lives of poor farmers and then standing cold, dead—altogether a sorry successor to that willing servant and noble friend of man, the Horse. Steinbeck has told magnificently the terrible story of the many people who are left out of the abstractions of long-view economists; we need more novelists to deal concretely with the tragic dislocations that are the price of industrial "progress." But he also creates his own abstractions, interrupts his story to write essays in competition with the economists; and his generalizations are sometimes shallow and sentimental. To attack one efficient machine in a machine civilization is simply futile. At that, Steinbeck falsifies even concrete, individual experience. Actually, men do not regard their machines as soulless; they fondle them as they do all their tools, keep them sleek, call them "beauties."

Unquestionably, science is dangerous. So are all adventures in thought, all dealings with the world; so is the whole experiment of civilization. In periods of crisis, moreover, all ideal instruments are abused. Men go to war equipped with the terrible weapons now provided by science; but they still go in a religious spirit, following the gleam, living the myth. At any rate the basic fact remains that science has come to stay and is in a position to give rather than take orders. If it has yet to become naturalized, it is nevertheless omnipresent as Jehovah never was, at once a spirit moving upon the face of the waters and a man walking in the garden in the cool of day. We therefore need to see just what it is, what it can and cannot do, and then to feel at home with it. We finally need to bring it into harmonious relation with the other values by which we live. For the split between our intellectual habits and our patterns of living, Hogben remarked, can easily be fatal to both.

2. WHAT SCIENCE IS

ROUGHLY stated, the scientific method is to go and look, and then look again. The most elaborate experiments and abstruse equations are designed to answer the simple question, "What are the facts?" Today this question seems so natural and obviously sensible that it is hard to understand how for centuries men could repeat Pliny's statement, that the blood of a goat would shatter a diamond, when a simple test would have disproved it. Yet it seems that they did not perform the test; and the explanation is that the basis of their thought was not empirical but "rational." Although Aristotle went to nature, he returned for authority to pure reason. He simply asserted that heavy bodies must fall faster than light ones, just as he asserted that planets move in circles because the circle is the only perfect figure. Hence Galileo's Pisa experiment marked a real revolution in thought. It marked, Dewey summarizes:

> a change from the qualitative to the quantitative or metric; from the heterogeneous to the homogeneous; from intrinsic form to relations; from esthetic harmonies to mathematical formulae; from contemplative enjoyment to active manipulation and control; from rest to change; from eternal objects to temporal sequence.

In this summary, science already begins to look strange to the plain man; and of course it is strange. Even as roughly stated, its method is still not generally applied to moral, political, or other problems. For science is not, strictly, "organized common sense." Common sense is not only much vaguer and more cocksure but in a way, curiously, more practical. It deals with the total concrete situation, takes life as it comes. Science always abstracts for a very limited purpose, makes up fictions. Especially in late years, it has left common sense far behind. When scientists try to speak the plain man's language, they tell him that the quantum theory may be understood by the analogy of a clock whose mechanism had vanished, leaving only the ticks, and that if he still doesn't understand, the point is that the universe is "not only queerer than we suppose, but queerer than we *can* suppose."

Yet science does remain simply a form of organized intelligence; to become oriented to it, we again do well to begin with the obvious. Although men talk as if the object of intelligence were the discovery and contemplation of eternal truths, actually they employ it chiefly to handle the new situations that are always arising even in a routine life. In daily experience they are continually experimenting, reconstructing, adjusting themselves to a continually changing environment; otherwise there could be no consciousness, no real experience at all. The scientific method is a systematic extension of this behavior. George H. Mead therefore described it as "only the evolutionary process grown self-conscious." Biologically, it is an advance in the natural direction: more differentiation, finer adaptation to environment, greater control over environment.

Similarly the basic interests of science, the concern with the "material" world, are not actually newfangled or alien. Men often feel that nature is hostile to them, at best very careless, at worst unfathomably cruel; in their philosophies they have represented it as a show of illusory or accidental appearances, in their religions as a mess of devil's pottage. Nevertheless they also feel a deep and constant kinship. They naturally personify the world about them and draw from it their metaphors for human life: they bud and bloom in youth, they ripen like fruit on the bough, they fall into the sere, the yellow leaf. The rhythms of nature are in their blood. Like poetry, science explores and articulates these relations; it realizes our rich heritage as children of this earth. Like Christian theology, moreover, it assumes that the heritage is lawful. Science grew out of the medieval faith that the world is orderly and rational, and that all happenings in it could be explained. Scientists now consider this a postulate, not a fact, and their explanations are usually offensive to orthodox theologians; nevertheless they have the same working faith as the theologians. Thus Newton could lay the foundations of the mechanistic universe in a spirit of extreme piety, and be applauded by other devout Christians; he was simply clarifying the ways of God to man. Thus agnostic scientists still admire all the evidence of uniformity, regularity, harmony in the universe. They admire the most wonderful of miracles, that there are not incessant miracles.

In other words, they are not really so inhuman as they are reputed to be. Whereas the man on the street sees only the gadgetry of science, intellectuals are prone to the other extreme of viewing it always in the abstract. They dwell upon its remorseless impersonality, the coldness of its truth; they forget its personal satisfactions, the imaginative value of its truths. For to scientists truth is indeed beauty. Mathematicians exclaim over the "elegance" of their demonstrations, Einstein delights in the "pre-established harmonies" that physicists discover, J. W. N. Sullivan is struck by the "astonishing beauty and symmetry" that Minkowski gave the theory of relativity by adding the notion of a four-dimensional continuum. On the other hand, they are displeased by unsightly gaps or bulges in their theory-patterns, dislike the messiness of quantum physics even when its theories seem to fit the facts. Their effort is always to get all their facts to fall into a shape, and their preference among theories, when the experimental test has yet to decide, appears to be determined chiefly by the esthetic quality of the shape. Thus Sullivan notes the comments of Einstein and Eddington on each other's attempt to reduce the laws of electromagnetism to geometry: Einstein said he simply did not "like" Eddington's theory, though he could not disprove it, and Eddington said Einstein's theory was a matter of "taste." Altogether, the generic motive of science is no doubt utilitarian—"service to mankind," if one likes more exalted terms; but the individual scientist, like the individual artist, does his work for the simple, unexalted reason that he likes it, and when it turns out right he feels a comparable lift and glow.

The simple answer to Ransom, then, is that science does "free the spirit." He has forced a narrow view of utility upon it, just as moralists and scientists often do upon art. Like thought itself, science has become a passion and a luxury. It follows the gleam, it stirs hopes too wildly dear. It is indeed often not utilitarian enough: science for science's sake is as much a cult as art for art's sake, and can carry one as far from the actualities of purposeful living. Yet this same passion calls out the plain answer to Roelofs. Science does produce saints. Not to go down the long list of heroes and martyrs, Mme. Curie will do as an example of simple, noble goodness. Such idealism is not itself scientific, to be sure, and may be called religious. Nevertheless the fact remains that science can inspire it without benefit of clergy.

This demonstration that even the scientist is human may seem inconsequential. It finally leads, however, to the heart of the problem of what science is. The recent developments in its philosophy may be summed up in precisely this recognition of the "human element," the human "standpoint" that is literally involved in all statements. Scientific laws are not chips off the old block Reality; as interpretations of sense impressions, they take after the human mind as well. All knowledge is a joint enterprise, an affair whose conditions are both inside and outside the organism. It is the offspring of the marriage of man and nature, a union in which the older partner may be expected to outlive the younger but which is indissoluble during the life of man.

This idea will concern us later on. Immediately, Einstein tells us how to understand the scientist's method: "Don't listen to his words, examine his achievements." Still better, watch him at work, examine the actual operations by which he gets his knowledge; and here an excellent guide is William H. George's *The Scientist in Action.* Whatever it may become in theory, George points out, a scientific fact is in practice an observation of coincidences. Although products of sensory impression, facts are impersonal in that they are independent of the judgment of any one man; they are statements of coincidences that can be observed under the same conditions by all men. The scientist can therefore gather and test them without bothering about such philosophical problems as whether there really is an external world; "real" is not an observable property. He does have to bother, however, with the problem of classifying and interpreting his facts, fitting them into patterns called theories and laws. The more comprehensive these are, the better he is pleased; but the most comprehensive is still tentative and does not "reduce by one the number of absolute truths to be discovered." Newton's great laws were patterns into which hitherto unconnected facts could be fitted; Einstein devised a different pattern that could accommodate all these and other facts; and we may expect that more inclusive but still different patterns will be devised by Zweistein, Dreistein, etc.

In other words, facts and figures do *not* speak for themselves. For all their stubbornness, they are accommodating enough to allow a number of different interpretations—and there are always enough of them around to support almost any theory. Moreover, the facts are not simply there, waiting in line to be discovered. The scientist selects from a host of possibilities, he looks *for* as well as *at,* he may accordingly *overlook*—as Grimaldi's experiments on the path of light were long neglected because they did not fit in with Newton's corpuscular theory. Hence the advance of science has not been automatic or really systematic, and it has not been in a straight line. Science is first of all the creation of scientists, who are also men with temperaments, special interests, predispositions. (Bertrand Russell has noted, for example, the divergent developments in animal psychology under Thorndike and Koehler: "Animals studied by Americans rush about frantically, with an incredible display of hustle and pep, and at last achieve the desired result by chance. Animals observed by Germans sit still and think, and at last evolve the solution out of their inner consciousness.") More significantly, science is the creation of a definite type of mentality, which has been interested in certain kinds of phenomena but notoriously indifferent to others, averse to the seeming "wild data." Most significantly, it is the creation of a culture, a society with special interests. Even physics, which seems wholly impersonal and autonomous, has been influenced by vested social interests. The concept of energy was developed to meet the manufacturers' need of a bookkeeping device, a way of measuring the efficiency of machines in units of work; in general there is an obvious correspondence between the long reign of classical mechanics and the needs of industry. Today, when science has developed a highly specialized technique, language, and subject matter of its own, it is still dependent upon the greater society for its privileges. It is the more profoundly a fashion of the times.

This view is not designed to humble or discredit the scientist. Rather it relieves him of the awful responsibility of speaking absolute truth. It stresses his continuity with the organic processes of evolution, the tremendous adventure of civilization, the vital needs and purposes of society; the scientist no more than the poet can afford the illusion that his activity is pure or priestly. It makes clearer the cultural pattern of science today: the concept itself of patterns, fields, organic wholes, which—as we shall see—has become important in all the sciences, and which parallels the collectivistic trend in the world of affairs. And it enables a more realistic approach, specifically, to the difficult issue of just where science properly begins and ends.

The popular notion is that science necessarily involves the use of instruments in a laboratory. Knowledge cannot be really scientific unless men have got it out of a test tube, taken an X-ray picture of it, or tried it out on some guinea pigs. Such methods are very well for dealings with sticks and stones, animal life, or the human body; but it follows that they cannot apply to the motions of mind or spirit. Laboratory workers themselves are often contemptuous of the social sciences, and of psychology when it leaves the laboratory and deals with such immeasurables as "consciousness" and "insight." They distrust any statement that cannot be put into an equation. And so the critic is warned off the sciences of man, which are naturally closest to his interests. He is left with the problem of determining just where, then, the sciences stop and the humanities begin, and just what use he can make of the power that has in any event so thoroughly made over the world in which the humanities have their being.

To begin with, there are important distinctions that should remain distinct. Some generous philosophers identify science with all disciplined thought, uniting all the humanities and the sciences in one big happy family. Thus Cassius J. Keyser defines science as any work that aims to establish by legitimate means a body of categorical propositions about the actual world; he therefore accepts as science the work of Plato and Aristotle—and blurs the fundamental difference between their thought and the thought of Galileo or Darwin. Moreover, there are important differences between the sciences. The physicist and the chemist have the adventitious advantage of large subsidies (capitalism has been a generous if not a disinterested patron) and now of relative freedom from personal prejudice or official interference; the psychologist and the sociologist are at any moment likely to tread on the corns of public opinion or get mixed up in some live social issue. But the former also have the intrinsic advantage of a subject matter that lends itself to the extremely helpful devices of mathematical measurement and controlled experiment. The experimental test is especially important, as the ultimate criterion for distinguishing scientific knowledge from philosophic speculation.

Nevertheless most distinguished scientists appear to agree with Max Planck, that from physics to sociology there is a continuous chain; and I can see no practical or logical reason for choosing to break the chain. On practical grounds, it would seem desirable to give science as much scope as possible, and not to discourage important social inquiries by verbal quibbles or qualms about their scientific chastity; it would seem foolish to demand complete, positive knowledge or none. On logical grounds, any sharp break in the chain is not only arbitrary but inconsistent with the basic scientific assumption of natural continuity. That the physical sciences are more objective and more exact than the sciences of man makes them neither more fundamental nor fundamentally different. The differences are in degree, not in kind.

Ultimately the unity of science lies in the logic, not the materials or the specific techniques of its inquiry. As formulated by Dewey in his monumental work, this is a logic of discovery and invention. Its forms are not a priori but postulational and operational; they are not absolute modes of pure reason but generalizations drawn from previous inquiry and liable to modification by subsequent discoveries. Indeed, scientists object to any theory, such as vitalism in biology, which is complete and therefore offers no possibility of advance; their curious objection, J. H. Woodger observes, is that it is *too* successful, *too* perfect. They demand that all theories live dangerously. But this experimental logic does not absolutely require the specific technique of laboratory experiment. It requires primarily that theories be so formulated as to leave room for future discoveries and almost certain modifications. It thereby exposes, indeed, the essential weakness of the sciences of man today, which is not so much the jungle growth of theory as the attitude toward this theory. As scientists, psychologists and sociologists are still very young, and like youngsters much too cocky—few physicists speak with quite the assurance of John B. Watson or Pareto. More specifically, they are seldom content with mere postulates and approximates; they set up some explanatory principle as necessary and sufficient, the one positive truth by which all the other little truths must be sired or certified. Yet their attitude is quite gratuitous. This very criticism of it implies that an experimental logic can be applied to these problems too.

"Wherever there is the slightest possibility of the human mind to *know*," wrote Karl Pearson, "there is a legitimate problem of science." If men have "known" all sorts of absurdities, there can be no question about a fact, strictly defined, and such facts are available in all spheres of interest. Observation, not measurement of coincidences is their criterion. If it is clearly more difficult to classify and interpret them in the sciences of man, it is not clearly impossible; important relations have already been established and systematically formulated. Students of the humanities who deny that there are fundamental laws in their province necessarily think in a way that presupposes such laws—else their thought would be pointless. In sum, only by divorcing human affairs from natural processes can they be shut off from scientific inquiry; and this ancient expedient disposes of the problem by creating two more.

3. THE USE AND ABUSE OF SCIENCE

THE practical uses and abuses of science are generally conspicuous enough. They have alike, however, confused the issue of its value. Men consider chiefly the *results* of scientific inquiry, and build their philosophies on current theories as if all thought labeled scientific were necessarily valid. They neglect the logic of inquiry, the *method* that has produced all the dazzling results and that will survive all the current theories. The disciples of science are therefore apt to become only another species of theologian. Still more superficial, however, is the view that science is useful only for menial work, the chores of society, and should be shut off from spiritual matters. This aristocratic division of labor is in effect a return to mumbo jumbo. Its advocates are trying to achieve ends without a realistic study of means, and ultimately they divorce ends and means. In placing mind or spirit in an unearthly realm of its own, they do not actually elevate it; they leave it up a creek.

To begin with, the findings of science are clearly relevant to spiritual interests. There can be no sound judgment of any values without a knowledge of actual conditions, the nature and possibilities of man and his environment; about such matters the scientist has a great deal to say. Grant that his primary interest is in processes, apart from all considerations of value, and that the critic's primary interest is in outcomes, in which considerations of value are central; still a knowledge of processes is always helpful in a study of outcomes.

Even more pertinent is the method of science. Theoretically it is depressing, because it assumes determinism. Actually it gives us freedom, freedom where we want it and can do something with it—freedom, as Dewey says, not apart from natural events but in and among them. It is the philosophers and theologians who have characteristically restricted both our thought and our action; setting out from fixed truths instead of hypotheses, they permitted less freedom than the experimental method, which may imply determinism but which remains *experimental.* Science, therefore, not only has proved an efficient instrument of human purposes but has created new purposes, liberated purposefulness from the ends arbitrarily fixed by ancient high priests of thought. Likewise its method is fundamentally more humane. The philosophers and theologians have characteristically set up absolute antitheses, of right and wrong, good and evil, and conceived their first duty to be the annihilation of most other thinkers. Scientists work more by assimilation, a reconciliation of opposites that are more apparent than real. Each new theory recaptures from a higher level the elements of truth in previous theories, and is in turn due to be included in a still higher synthesis; Newton's great contributions remain solid after Einstein devised a larger pattern. In short, despite its supposed inhumanity, science exemplifies the humanistic ideal.

Also pertinent here is the moral value of science, which is so apparent that it is often overlooked. In no other human activity is a higher premium placed upon truthfulness, or is the ideal of truthfulness more fully realized. Success in science is possible only on these terms. The eminent statesman or businessman may be unscrupulous, and indeed cannot afford too many or too fine scruples; the passionate sincerity of artists and other intellectuals may still be warped by wishful preferences, and their fervor and eloquence come by more easily because of the wishfulness; but no scientist can ever hope to succeed by misrepresenting, or by remolding according to his heart's desire. (Paul Kammerer, the well-known biologist who did fake his results, was driven to suicide by the disgrace of exposure.) His hard-won success, moreover, cannot be exploited for private gain, at least if he is to retain his professional standing. Outside his profession, needless to add, the scientist is no paragon of selfless virtue and wisdom, and within it he will cherish his brain child, hold out for his own theory as long as the facts permit. Yet the scientific community as a whole is the most impressive example history has yet known of a disinterested, coöperative enterprise, international in scope, directed toward impersonal ends and by impersonal standards. It is the most impressive demonstration of the actual possibility of "supra-personal, supra-partisan, supra-racial standards and values."

In such ways, to repeat, science can free the spirit. So can its tremendous revelations, which have also been considered alien to the human spirit. Astronomy is the striking example. Men are often depressed by the immensities it reveals: How insignificant is man, how transient his life in the perspective of the stars! Such reflections may induce a becoming humility. But neither can man lick a gorilla, or outrun an antelope, and this popular habit of measuring him by astronomical coördinates is also pointless. We may as reasonably look down on the stars, for they cannot contemplate either themselves or us. Meanwhile astronomy has other meanings for us. The stars were the first clear evidence of uniformity, regularity, harmony in the world; man might never have achieved any science at all, Poincaré remarked, had clouds always hid them. They have as well a clear imaginative or spiritual value. The spectacle lifts us out of ourselves; and if it shows how little man is, it also shows how great is his spirit, which can enjoy the immense harmony and embrace the immense mystery.

In general, science has unquestionably widened and enriched the whole context of thought, the whole background of immediate experience. The issue of whether or not to use its findings is in the end academic. Try as we will, we cannot escape its influence. We cannot walk abroad except in its shadow; it is in the air we breathe, the language we use as unconsciously. The student of esthetics, for example, may rebel against the excesses of "psychological" investigation and try once more to take his beauty straight. Nevertheless his main terms—imagination, emotion, sensation, will, intuition, expression, etc.—are charged with some kind of psychological theory. That it is usually a vague or dubious theory only emphasizes the need of his realizing what he is about. And the moment he leaves esthetics to generalize about the relation of art to other interests and purposes, he is up to his neck in matters about which scientists have had much to say. He is perforce a kind of psychologist and social scientist. He might better be an informed one.

Unhappily, however, we cannot escape the influence of pseudoscience either; and so we must go back to the beginning. The extraordinary success of scientific inquiry has also stimulated a busy shopping around to pick up some scientific dress at bargain rates. The most obvious bargains are in technical jargon, which may then clutter up not only a man's prose style but his thinking; he is apt to think that he has solved a problem when he has merely located it, as in the Unconscious, or given it a new name, such as "sublimation." But the general reason for all the awed misunderstanding is again that laymen usually pick up only the latest findings, not the logic of inquiry. With its elaborate system of fictions and symbols, which may or may not stand for something "real," science might be defined as madness with a method. To miss the method is to be left simply with madness.

Perhaps the most common and dangerous fallacy is the assumption that science tells us "all that matters." Max Eastman stated it baldly when he insisted that poets must stop "trying to tell us anything about life," because as poets they know nothing about it. This is a hard thing to tell an earnest man. It is also a very silly thing, and should embarrass scientists as much as it distresses poets; the exclusive right to say something about life is a terrible responsibility. Yet scientists, too, often talk carelessly in this way. Physicists will say that a table is "really" only a swarm of particles, or biochemists that man is "nothing but" a parcel of chemicals worth about a dollar. All that is of no value for their limited purposes—the distinctive qualities of human experience, the extraordinarily rich variety of sensation and emotion in a world of shape and color, fruit and flower, dream and song—they may then dismiss as merely "subjective," beneath the dignity of knowledge. They forget that they are always abstracting, and that their abstractions are also fictions. "It is the Nemesis of the struggle for exactitude by the man of science," writes H. S. Jennings, "that it leads him to present a mutilated, merely fractional account of the world as a true and complete picture."

The greatest victories of science have in a sense been cheap victories, for they have been won by a careful refusal to engage the more difficult, certainly no less important problems of ethics, politics, and the humanities generally. The physicist rules out all that he cannot rule, reduces what is left to the simplest possible terms. It is an excellent policy—for his purposes. But he may then forget that what cannot be measured or calculated may nevertheless be immeasurably important, have incalculable consequences, and that the terms of life as it is lived are not simple. Take even the unanswerable questions, the metaphysical questions that it is now fashionable to dismiss as meaningless. "All problems are artificial," Dewey wrote in a careless moment, "which do not grow, even if indirectly, out of the conditions under which life . . . is carried on." Roelofs answers rightly that *all* problems grow out of these conditions. Where else could they come from? Metaphysical speculation may be unproductive of strict knowledge, and is often a dangerous nuisance because it is mistaken for knowledge. Its high imaginings may also be a source of inspiration and strength. In either event the fact remains that men naturally wonder about the ultimate mysteries, give rich meaning to meaningless questions, and behave in accordance with the answers they have given the unanswerable. And so they must, for many indispensable meanings and purposes outrun all evidence.

At most, science gives us certain directions for dealing with certain things. Its abstractions are like paper money, facilitating commerce with actuality so that every transaction need not be an exchange of cows. Its abstractions are a necessary evil; and they become simply an evil if men hoard them for their own sake, forget their instrumental function and their arbitrary value. But it is especially important for the literary critic to remember that science never tells us everything about anything—that, in fact, it tells us very little about any *thing*. It leaves out of its abstractions all the unique particulars, the concrete qualities, the substantial body of direct experience that is the primary subject matter of art. The scientist's apple suggested the law of gravitation and as happily obeys the laws of biology. The apple of our experience is a unique object that may please the palate and the eye, evoke the emotions appropriate to autumn, symbolize the plight of the unemployed, enable an archer to become a national hero. Our apple is no less significant than its better behaved symbol.

Similarly the critic in particular needs to guard against the confusion commonly inspired by scientific analysis. Analysis can reduce all human behavior to its biological origins and these in turn to their physical basis; then men believe that they have completely "explained" behavior, proved that it is "simply" physical. A great deal of emotion has been generated by the elementary fallacy pointed out by Broad: because B grows out of A and C grows out of B, C is nothing but A in disguise. Scientists know that organic life is qualitatively different from inorganic matter, the conscious life of man qualitatively different from other forms of life; they may nevertheless forget that what matters most to us as men is not the common beginnings but the unique ends. Critics fall into the same fallacy when they confine science to the "economic impulse" because it is rooted there. But they also fall into it when they describe art as "merely" a compensation for some psychic maladjustment, or "essentially" an expression of class interest. They confuse the conditions of something with its cause, and both with the thing itself.

The gist of all these extravagances is that science has suffered as much from its disciples as has any faith. For it is indeed a faith, and often a very naïve one. Apart from the gross worship of time-saving and time-killing devices as ends in themselves, the disciples have displayed the artless optimism of prophets and crusaders through the centuries, based on the illusion that men always welcome light and once brought to see the truth will act upon it. They have displayed as well the frequent arrogance, intolerance, and inhumanity of the prophets and crusaders. Thomas Huxley, who declared that faith is "the one unpardonable sin," had through all his career an absolute faith in the religion of science, accepting as gospel not only its method but its current assumptions, and on the basis of these questionable assumptions erecting an often harsh philosophy. Others have been contemptuous of values other than their own devotion to verifiable knowledge, contemptuous even of the vital needs they were presumably serving. Until recently, "purpose" and "value" were the most horrid words a scientist could hear.

All in all, science is a means to a better life on earth, but cannot alone secure this end. The most that one can argue for it is that it is an indispensable means, the best investment for hope. Yet this is a sound argument; and with it I should end. Philosophy, religion, and art provide valuable experiences and suggest valuable ways of dealing with the world; science provides the most certain knowledge, the most certain means of generating more knowledge, the most certain means of manipulating this knowledge to control the natural world. Hence the necessary postscript even to preliminary discussion is that scientists now tend increasingly to think scientifically about their own activity. They are examining the implications and facing the responsibilities of the enormous power they wield. They are subjecting science to the same ruthless analysis that religion has had to endure since the eighteenth century, conducting a searching inquiry into their basic assumptions and their logic, systematically formulating the philosophy that in their naïve, exuberant youth they did not realize was a philosophy. Poincaré, Whitehead, Planck, Einstein, Jennings, Woodger, Koehler, Mannheim—such men offer a profounder criticism of science than one gets from most thinkers who are jealously defending their values from its encroachment. They are less ignorant of their ignorance.

IV

THE PHILOSOPHY OF PHYSICS

But yet the greatest certainty, advanc'd from supposal, is still but Hypothetical. So that we may affirm, things are thus and thus, according to the Principles we have espoused: But we strangely forget ourselves, when we plead a necessity of their being so in Nature, and an Impossibility of their being otherwise. JOSEPH GLANVILL.

1. THE REVOLUTION IN PHYSICS

"SEE Mystery to Mathematics fly!" wrote Pope in a simpler age. Today the layman who attempts to follow the flight through the probability waves in the time-space continuum is apt to appreciate the blessed old mysteries. The chief trouble is that he really wants to *see*. He lives in a common-sense world full of material things and uses a language full of nouns, which by definition are "substantives" or names of these things. Stubbornly he asks, What *is* an electromagnetic field? what *is* a line of force? To such questions physicists are blandly indifferent. They still tolerate "matter" but only on sufferance; many look forward to a day in which they can dispense with this "theoretical construction" and explain everything in terms of "field"—although this too is not necessarily "real." In general, they do not care what their symbols "stand for" so long as they can get handier equations. Space has the physical property of transmitting electromagnetic waves, Einstein tells us, and we should not "bother too much about the meaning of this statement."

Yet one need not fly all the way to mathematics to catch the main idea of what is going on—the specific equations will probably have been revised before he has got to them anyway. The permanent contribution of modern physics lies in its new base of operations. Between the calculations of Newton and Einstein there is only a slight difference; for all ordinary purposes one can still measure things in the old-time way. Between the implications of their theories, however, there is an enormous difference. And this difference, in which lies the profoundest revolution of an age noisy with revolutions, the layman can grasp.

In its main outlines, the story of what has happened is familiar enough. Classical physics explained everything in terms of matter and motion in Euclidean space, running the ancient forms of Being by as immutable a clockwork mechanism. From the beginning there were troublesome fictions, such as the apparently jelly-like "ether" that transmitted light but somehow offered no resistance to the wheeling spheres; yet the whole scheme was built solidly on common sense, and it seemed to work beautifully. The first principles of physics were accordingly regarded as a priori, its framework and method as inevitable and inalterable. Jealous philosophers kept raising questions, asking how our senses enabled us to be so intimate with the little lumps of matter and how these lumps kept pushing one another around, but the very success of scientists resulted from their indifference to such questions, their naïve acceptance of a faith without bothering to explain or justify it. In time, however, they accumulated more and more experimental data—especially regarding electricity—that could not be explained satisfactorily by their mechanistic concepts. Clerk Maxwell's brilliant electromagnetic theory of light did *not* have a mechanical basis, yet it also seemed to work. Hence their very successes finally forced scientists to look to their faith. And even as Renan was exclaiming, "The world today has no more mysteries!" this tidy world was crumbling.

The story may conveniently begin with one of the greatest but least known of Russian revolutionists, the mathematician Lobachevsky. Euclid's axiom, that through a given point only one parallel could be drawn to a given line, had been considered a fact of nature, plain to the eye. But in 1826 Lobachevsky, as if just for the hell of it, denied this self-evident truth—and built up a whole consistent geometry on the assumption that *more* than one parallel could be drawn through this point. Then another mathematician constructed a new geometry on the assumption that *no* such parallel could be drawn; and thus there developed flocks of geometries. These have in turn proved useful to physicists. (In quantum theory, I gather, profitable use has been made of an arithmetic in which 2 times 3 does not equal 3 times 2.) Accordingly they too changed their base of operations. Einstein challenged the axiom of simultaneity, that two events can happen in *different* places at the *same* time, and thereby developed his theory of relativity. All along the line physicists have arrived at more satisfactory interpretations of experimental facts by scrapping self-evident truths, breaking the laws of thought—by a systematic exploitation, as it were, of the nonsense that the eighteenth century had triumphantly eliminated. They pride themselves chiefly on the possibility of asking still more preposterous questions and getting still more preposterous answers.

The important contribution of modern physics, then, is not the particular nonsense that it will erect into the truth of tomorrow. It is the junking of Newton's absolutes, the breaking up of his or any other fixed frame of reference, the overthrow of the totalitarian state in the world of thought and the establishment of a democracy in which all hypotheses are freely elected. The revolution might be summarized as the triumph of the postulate over the axiom. An axiom is something self-evident, fixed, unquestioned. A postulate is something assumed, to be tested for its usefulness—not a law laid down by God but a logical fiction consciously invented by man. "No one can say," declared Descartes of the properties of triangles, "that I have invented or imagined them"; mathematicians and scientists now say just this. We must be very careful, writes P. W. Bridgman, that "our present experience does not exact hostages of the future." This might seem too squeamish a concern for posterity, which can be trusted to take care of itself; but scientists are concerned chiefly with their own experience. Newton's assumption of Absolute Time had restricted their outlook and course of action for some two hundred years.

To come to the more specific concepts, it is common news that "matter" is no longer the inert, grossly "material" stuff of old. Physicists now represent objects as processes or events, trace dynamic patterns of an intricacy and subtlety that make the traditional operations of spirit seem crude. The most familiar element in our experience, matter has become more and more elusive, mysterious, incomprehensible. Further, the scientific definition of it is generally conceived as an idealization—a convenient formula, in Bertrand Russell's words, "for describing what happens where it isn't." Strictly, physicists do not know what they are talking about. They do not know what anything *is;* they tell us only what something *does*. Their descriptions are not photographs but ordnance maps for future operations. In a sense, accordingly, they do not so much uncover truth as create it. In *The Evolution of Physics* Einstein talks constantly of the "important invention" of the electromagnetic field and all the other realities "created by modern physics." He rejoices in the new concepts because they have enabled us "to create a more subtle reality"; he would therefore drop them instantly for concepts that made possible a still fancier reality. In other words, the reality known to man is not immutable.

It follows that scientific "laws" are not categorical imperatives. As Karl Pearson pointed out, they are shorthand descriptions of nature and cannot be said to *rule* it. That nature appears to obey them proves nothing, for they were invented for just that purpose; when more experimental returns come in, nature will obey some new, perhaps quite different laws. To give a new twist to the old religious argument, law indeed implies a Law-giver—who is Man. At any moment, moreover, there are various conceivable ways of interpreting the experimental data. Physicists always prefer the widest possible generalization and the simplest possible formula, seek to break nature down into as few elements and laws as they can; but this procedure is a convenient method, not an absolute necessity. "Nature is pleased with simplicity," Newton wrote, and certainly men are pleased with it; but of nature we cannot be sure. The latest investigations of the subatomic world suggest to physicists as well as laymen that it may be complex beyond the dreams of Marcel Proust.

Implied in these statements is again the human "standpoint." Alfred Korzybski makes out, roughly, three periods in the history of thought: the Greek period, metaphysical and idealistic, in which emphasis was primarily on the observer; the scientific period, semiempirical and materialistic, in which emphasis was primarily on the thing observed; and the period now dawning, in which knowledge is a transaction between the observer and the observed. On the submicroscopic level, quantum physicists are confirming Coleridge's suspicion of the mechanistic assumption of inert matter that can be observed without being disturbed. The quantum of energy leaving the electron and hitting the observer's eye can be measured only by a new observation, which in turn affects the electron; we cannot actually peep into the private life of the electron. On the cosmic level, Einstein has assumed that absolute time is as meaningless as absolute length or absolute cheese, and that all possible measurements of time and space necessarily involve our position. As Bridgman says, he seized on "the act of the observer as the essence of the situation." The world of Planck and Einstein seems strange, indeed, precisely because man cannot be left out of it. He is not only the most intricate but an indispensable piece of apparatus.

To admit the importance of the observer is also, however, to admit that approximateness is a necessary condition of human knowledge, not merely a matter of imperfect instruments. Uncertainty or mere probability has therefore been erected into a scientific principle. In classical physics one could, in theory, predict with absolute accuracy the future course of any bit of matter if one knew its position and velocity and the forces acting upon it. Heisenberg's Principle of Indeterminacy states that because of the very nature of things we cannot possibly know *both* the position and the velocity of any bit, and that the more precisely we determine the one, the less we must know about the other. Furthermore, in quantum physics there are no laws for the behavior of an individual particle but only statistical averages, "bookkeeping laws," for the behavior of the whole crowd. The physicist cannot even in theory make a definite appointment with a particular electron; he can state merely the probability of where and when it will turn up. He knows that in a certain period approximately so many radium atoms will disintegrate, but he does not know which ones are doomed, or why, or precisely when they will meet their fate. Hence another striking paradox: in this view the lawfulness of nature may be rooted in lawlessness, a very high degree of uniformity resulting only because billions of coins are tossed. What the physicist can state with mathematical exactness is the limits of the *in*exactness of his calculations; as in all operations of pure chance on a large scale, there is a predictable, measurable degree of *in*accuracy.

Now many physicists are dissatisfied with these theories and consider them mere stopgaps. Statistical methods, Bridgman observes, are generally used either to conceal vast ignorance or to simplify vast confusion. Such messy, haphazard behavior of electrons distresses men brought up on law and order; they accordingly look forward to a day when subtler legislation will again induce all the electrons to keep their appointments like little gentlemen. For what is involved here is the fundamental principle of causation. Some physicists regard the surrender of strict causal laws as a threat to the integrity, even the possibility of science. Others agree with Erwin Schroedinger, that cause-and-effect is mere "mechanico-morphism" and should be scrapped with other primitive habits of thought. Still others, such as Einstein and Planck, occupy the middle ground, believing that the traditional formulation of the causal principle is rough and superficial, but that the principle itself is still indispensable to science.

In the face of such distinguished disagreement, it would be brash of a layman to settle the issue. Yet he may wonder whether the concept of cause actually has been given up; the very physicists who argue against it say that we must give it up *because* of such and such facts. What is under fire, at any rate, appears to be only a particular kind of causation, the mechanical concept of classical physics. This was an outgrowth of the common-sense notion of cause as an external force—a notion rooted in our own experience of pushing and pulling, which W. H. George describes as "the triumph of muscle over mind." (Thus the peasant to whom the steam engine was explained asked to see the horse that pulled the locomotive.) The point of the new concepts in physics, however, is that gravitation, for instance, is not a physical something that *makes* apples fall. It is a formula, a concise way of saying that all apples *do* fall, and of linking this fact with other regular sequences. Cause-and-effect may therefore be considered a tautology, a restatement of the observed correlations and uniformities in nature. But however they are described, the important thing is that we can and do make out uniform sequences. The Principle of Indeterminacy applies only to our present *descriptions* of what goes on inside the atom, sets a limit to our possible observation. It does not necessarily apply to the *behavior* of the atom, much less of the stars, or destroy the fundamental assumption of continuity. Whatever they think, in their actual operations physicists continue to bank on continuity and regularity.

Hence the layman may safely leave this problem to the experts. It is significant and healthy that such questions are being raised, in view of all the "necessities of thought" that have unnecessarily hampered it, and the answer to questions so stated will not seriously affect the nature of our knowledge as it is now conceived. Whatever strict causal laws may be invented will for the physicists still be working hypotheses, not final truths. "As a matter of fact," wrote Max Planck, "we have no means whatsoever of proving or disproving the existence of causation in the external world of nature." And in any event the causes assigned are always relative and arbitrary, not absolute and complete, for they never have a beginning or end. The scientist stops at some point, behind which one can always go in search of still larger or deeper causes—the cause, say, of gravitation. Thus John Smith explains a noise by saying that little Willie just smashed a plate, and he disposes of the problem by spanking Willie. He could also make this cause the beginning of an endless analysis, leading through the laws of sound and the psychology of Willie to the whole content of human knowledge, the whole history of the human race. Any single event involves the entire system in which it takes place. Finally one is asking, what is the "cause" of the universe?

Strictly, such questions are meaningless; and they have tormented men for ages. If nature abhors a vacuum, then men are indeed children of nature. They must somehow fill in all the empty spaces in their picture of the world—just as the anguished Victorians could not bear the sight of a blank wall and cluttered up their rooms with bric-a-brac. The idea of an infinite, eternal universe has always troubled them, for they must have a beginning and an end; the idea of a finite universe has troubled them no less, for it leaves the imponderable emptiness beyond the borders of space and before the beginning of time. They then try to explain the inexplicable by giving it a capitalized name—the First Cause, the Prime Mover, Fate, God. Somehow they must provide a mechanism, find a reason why, justify or dignify what simply *is*. And modern science is distinguished by the calm acceptance of empty spaces, the calm awareness of the meaningless question—with the realization that to call a question meaningless is to make a significant statement about nature and the operations of the human mind. "What?" and "Why?" may stimulate scientific research, but they do not, strictly, constitute its subject matter. Its spirit is a thorough-going pragmatism.

This attitude results in part from the assumption that process and energy, not matter, is the fundamental fact; if one thinks of existence as activity rather than being, he is less likely to ask how the universe "came into being." But it results as well from the realization that pure reason cannot take us to the heart of reality. If all flubjubs are dingbats and this is a flubjub, then it must also be a dingbat. This syllogism, Ogden and Richards point out, carries absolute conviction; but it proves nothing about the existence of dingbats. Formal logic can never prove the truth of its premises. Man can never be certain that his logic is the logic of things, that the scheme of science represents nature completely and represents nothing else. For all human purposes man is in fact the measure of the universe; but nobody knows exactly what it is he is measuring. In the past, scientists forced on nature the limitations of the human mind, identifying their picture of reality with reality itself. Now physicists recognize that this was but a subtler form of the ancient anthropomorphic habit; man was simple creating nature in his own image. So, indeed, he must if he is to deal with it at all. But for efficient dealings he must also be aware of what he is doing.

2. THE HAZARDS OF IMPLICATION

In attempting to draw out the implications of the new concepts in physics, one has first to forego a deal of pleasing phantasy. Literary men have been inspired particularly by the Fourth Dimension: a nominal definition or mere convention of measurement that has become a symbol of all the magic of science. Thus we have J. B. Priestley (after J. W. Dunne) romping along this dimension into a new kind of Immortal Reality, reveling in a new kind of bed-time story, in the delusion that Einstein has paid for his ticket. One might remark that Priestley is a piker; if he likes extra dimensions he should turn to quantum physics, where the possible positions of just ten bits of matter form a thirty-dimensional continuum—and so he could go on indefinitely in this three-for-one ratio. But these ingenuous fables are harmless only when they are taken for fable.

More dangerous is the cocktail chatter about relativity and uncertainty, the giddy skepticism about the reliability of any human knowledge. In physics, the theory of relativity has not destroyed all constants but introduced more fundamental constants, made possible still more exact measurements of time-space relations. It does not limit these relations to our *consciousness,* make them a purely subjective affair; it refers them to the velocity of the physical system we inhabit, makes them still more objective. Neither does the introduction of a principle of uncertainty or mere probability imply that all statements are equally uncertain or all events equally probable. If the behavior of individual electrons now seems haphazard, it is still a safe bet that a flock of them will not suddenly quit a chunk of iron and become a toad. Thinkers who are most sure that nothing is sure, that all is merely relative and subjective, nevertheless smuggle in some objective standard of knowledge or truth; for how, otherwise, could they say that other inferences are mistaken—or even know that they are *mere* inferences? And so with the romantic misinterpretation of the scientific world of ceaseless process: all is flux, all is whirl. To the scientist, change emphasizes continuity; he does not deny constants and uniformities but simply looks for them in relations and processes instead of essences or lumps. Change itself is inconceivable without reference to something constant.

Somewhat more complicated, however, is the issue of determinism that is also involved in the new principle of uncertainty. The layman can easily share the scientist's indifference to the private life and fate of an electron, for he can see the analogy with human society, in which such laws as appear to govern the group do not necessarily govern the behavior of any one individual. He may also rejoice that a rigid, mechanical determinism is no longer the ruling dogma of practical physics. And he may appreciate Poincaré's remark about the "laws" of probability, that everybody believes them because observers accept them as principles of mathematics and mathematicians accept them as facts of observation. Yet observation makes clear that there are not only probabilities but stringent compulsions. If scientific laws are human formulations, not inviolable commands, they are nevertheless formulations of *facts,* they refer to inviolable conditions—as will be found by the apostle of indeterminacy who jumps off a cliff or tries to run his head through a stone wall. Outside the atom determinism still does hold, the behavior of objects we can get our hands on still is predictable, and our practical purpose is still to locate all the stop and go signs in nature. Those who delight in the apparently haphazard, irresponsible behavior of the electron are apt to end in actually haphazard, irresponsible thinking. Short of this they may forget the great value of the disagreeable principle of determinism. It is as indispensable for moral wisdom as for practical business; in our relations with men as with nature we have first to obey in order to command; and at worst, as a homely philosopher has said, the notion that we are being pushed from behind may keep us always aware that we have behinds. Altogether, we may summarize the issue at this point by saying that the principle of determinism has not actually been abandoned in science, but that as now formulated it is less frightening. It is no longer a doctrine of mechanical predestination; it is a description of invariant sequences and relations—of the orderliness in nature that alone makes intelligent behavior possible.

All this is to repeat that too many literary men use science only as a taxi to take them quickly wherever they want to go. A further trouble, however, is that some scientists are all too eager to give them a ride—especially if they are going to the transcendental. Eminent physicists seem to have a natural inclination to idealism, and the rarefied materials they now work with leave them wide open to intimations of immortality. From the realization that their world of protons and electrons is the product of human consciousness, Eddington, Jeans, and others have leaped to the conclusion that the world "must really be the stuff of our own consciousness," a "spiritual reality," a "world of pure thought." "The final truth about a phenomenon," writes Jeans, "resides in the mathematical description of it"; and as this key to the universe is an invention of the human mind, the way is eased to the old notion of the human mind sharing in the processes and purposes of a divine mind, working out the equations of the Great Mathematician. This is a flattering notion, especially to mathematicians. But these men forget that they themselves are the authors of this agreeable universe, and that their finished product is essentially what they had in the first place—the world of their own consciousness. As C. E. M. Joad points out, they confuse an approximate description of the world with a complete definition of it, the nature of our knowledge of the world with the nature of the world itself. They divorce themselves from the products of their own minds, project these into the external world, and then are struck with wonder when they find them again. It is not surprising that they should discover a Great Mathematician in nature, for they have planted Him there. It is surprising that they should forget how ancient is this custom of creating God in one's own image.

The exploitation of these fond hopes may provide a job for some unemployed religious emotion. It has no value for scientific purposes, however, and it may obscure these purposes; the talk about "the final truth of a phenomenon" and what reality must "really" be misses the whole point of the new concepts in physics. By now, to be sure, this neo-idealism has been pretty well exploded by other scientists. Yet all the most advanced scientific thought is fraught with similar dangers, dangers to which the most brilliant mathematicians and logicians—such men as Alfred Whitehead and Bertrand Russell—are the most likely to succumb. Because of the refined materials with which they are now dealing, as well as the reaction against the naïve materialism of the nineteenth century, they tend to empty physics of all solid substance, regard it as pure mind stuff of another kind. "Only mathematics and mathematical logic," Russell declares, "can say as little as the physicist means to say." Here, indeed, it might seem especially presumptuous for the layman to offer criticism. Nevertheless his inability to follow these men into the higher mathematics and the farther reaches of symbolic logic can be a real advantage. If he cannot see what they are up to there, he can see that they often fail to return. He can see that they tend to lose touch with the empirical world precisely as did the medieval schoolmen.

Even to specialists, the nature of mathematical entities seems to be not altogether clear. Though these entities originated in concrete experience, the practical operations of measurement, they have undergone so elaborate and independent a development that Jeans and others are amazed when it is found that the end-products can still be applied to real events. At any rate, physicists can now express so satisfactorily in equations the formal relations between physical elements that they tend to forget the elements themselves. They forget what Dewey constantly hammers away at, that the final test of all scientific theory is not mathematical or logical but *experimental.* Although exercises in symbolic logic might conceivably carry us almost anywhere, where they carry physics is determined by experimental facts; although Einstein's theory of relativity was a brilliant mathematical achievement, it was accepted by physicists only after it was verified by further observations. Physics ends where it begins, not in a self-contained world of symbols but among natural events. All the pointer-readings that Eddington regards as a closed circle point to these events, in the last analysis to something we can handle or look at. In short, physics is still about the physical world, and mathematics is only the language with which it expresses the doings in this world.

This is to simplify epistemological problems, of course, and ride roughshod over some of the most subtle, original, and daring speculation of our day. But there is a real need of returning to a homespun pragmatic attitude—the attitude of the scientist himself in his laboratory—for these speculations have become so finespun that they are simply embarrassing science. Russell himself has pointed to their serious consequences: "While science as the pursuit of power becomes increasingly triumphant, science as the pursuit of truth is being killed by a skepticism which the skill of the men of science has generated." It is the consequence of separating theories of knowledge from actual practice; and this in turn intensifies our main problem, the split between our intellectual habits and our habits of living.

Nevertheless the most hardheaded men of science, curiously, have by different means contributed to the same effect. In his eagerness to get to "spiritual stuff," Eddington stresses the "unreality" of the pointer-readings until one might gather that science gave us no knowledge of the world at all but was mere play of mind. In the violence of their reaction against this spiritual stuff and all notions of Eternal Truth, Eric T. Bell and other realists are betrayed into the same false emphasis. Truth, Bell insists, is simply the elaborate tautologies that come out of the arbitrary assumptions dumped into the hopper of a logic-grinder; "the whole business from start to finish is a purely human activity." And so, within limits, it is. But the limits are imposed by nature as well as human nature, the business is not laissez faire. The physicist's activity is so arduous because he meets a brute resistance, can never change the rules to suit himself, must always make his laws work. His inquiry is *controlled* because something is given, everything is not merely taken. "That we do not *construct* the external world to suit our own ends in the pursuit of science," declares Max Planck, "but that . . . the external world *forces* itself upon our recognition with its own elemental power, is a point which ought to be categorically asserted again and again in these positivistic times." The very revolution in physics proves his point. It did not result because scientists freely decided to grind out some simpler or fancier tautologies, design some more elegant patterns; as a logical system, Newtonian physics is indeed more satisfying than contemporary physics. The revolution was forced by experimental facts that would not jibe with the old assumptions.

Hence all the terms I quoted in my exposition need careful

qualification. The physicist "invents" theories, "creates" causes; but he does not invent and create at random, he cannot play fast and loose with nature. His "laws" are convenient descriptions, but they are descriptions of an actual lawfulness; that night follows day and all apples fall down are not arbitrary assumptions. His whole scheme is not a literal copy of reality, but neither is it mere fiction or artifact; otherwise one might well ask why he should invent so slovenly a fiction as the quantum theory, so disagreeable a fiction as bacteria. In the language of common sense, our knowledge is approximate but it is an approximate to *something,* it is a knowledge of appearances but these are appearances of *something;* the practical scientist, like the plain man, has banged his head against too many facts to assume otherwise. And he knows that he knows something about this something, because he can do something about it. This is indeed a reasonable conclusion from his logical premises of continuity, interaction, the mind as natural product and not pure spirit. The "realistic" assumption—the assumption, as Bradley described it, that "the Real sits apart, that it keeps state by itself and does not descend into phenomena"—is at bottom a variant of pure idealism.*

I have labored this point because it seems to me fundamental for preserving the extremely important contribution of modern physics, which has been a solid gain, a genuine *advance* in knowl-

* As an extreme example of such realism, I cite Bertrand Russell's statement of "What I Believe" for the *Nation* (April 29, 1931). The necessity of condensing a whole philosophy of life into an article, as well as the exaltedness of telling the whole world what he believes, will naturally cause a man to simplify and overstate; but the direction of Russell's overstatement was significant. He declared that "order, unity, and continuity are human inventions"; such orderliness as we appear to find in nature "seems to be due to our own passion for pigeonholes." He repeated the familiar story of how modern physics "has plunged the world into unreason and unreality." But how or why, then, does physics do this? Why should Russell recommend the scientific method—"the habit of basing opinions upon evidence"? Evidence of *what?* Though he added that "human invention can, within limits, be made to prevail in our human world," he apparently regarded this as a curious coincidence, and did not say where or what this human world is. In fact, he cut it loose in a sentence: he said flatly that the two questions, about the nature of the world and about how human life should be conducted, "have nothing to do with each other." Here Russell went far to the right of most idealists and supernaturalists; they at least suggest some kind of relation between the two worlds. Logically, he divested science of all authority, except perhaps for those who share its consuming passion for pigeonholes. And though today he occupies a different position, as stated in *An Inquiry into Meaning and Truth,* the point remains that he has long been distracted in this inquiry by artificial problems.

edge. One should be able to reject Newton's absolutes without rejecting all objective knowledge and becoming a scientific atheist; for a positive atheism is as dogmatic as religion. Meanwhile we must be able to distinguish between more and less real; if the propositions of science have only a relative truth, they are always absolutely truer than the proposition that the moon is made of green cheese. We need to protect science from the demands for orthodoxy, now coming from revolutionary as well as religious quarters—demands that would be reasonable enough if scientific truth were merely a commodity manufactured for human convenience. We need to shoo off all the genteel thinkers who have been pleased to repeat that science too is a myth, and then to add that it has no more validity than the myths of poetry and religion—that as an upstart, indeed, it has less authority. We need, generally, to dispel the fog of obscurantism that has settled about the new concepts in physics.

On much the same grounds, however, we also need to guard against the logical positivists who will accept none but exact operational meanings, and who are far more concerned about syntax than about experience. Their favorite epithet, "unreal," is not a logically precise but a strategic term, and operationally a rather ineffective strategy; nothing disappears when it is called unreal. But they also do positive harm. With their contempt of all unverifiable speculation and belief, they are another species of the intellectual puritans who infest this age. Puritanically they deny the flesh of knowledge, insist on absolute separations, seek to narrow and rigidify science just as their ancestors narrowed and rigidified religion. In their reaction against the inveterate metaphysical habit, they forget that only the Devil now knows just what is metaphysical. All the operations by which a scientist gathers, expresses, and interprets his facts are enveloped in a haze. Throughout the history of science, very broad and speculative hypotheses—hypotheses at the time incapable of verification and never verified in their original form—have altered operational meanings and produced new ones. Today, with all the model atoms and fields and waves that we are warned not to regard as necessarily real, it is even harder to draw a fast line between physical and metaphysical, meaningful and meaningless. The final questions of what and why, of the essential nature and ultimate purpose of the universe, may safely be ruled out, for science does not suggest the slightest possibility of experimental inquiry, and philosophy offers only a shambles of irresistible arguments colliding with immovable thinkers. Nevertheless positivist physicists once considered meaningless the question of the weight of a single atom—a quantity now known. "The truth or falsity of an idea and the question whether it has a definite meaning is relatively unimportant," concludes Planck; "what matters is that it shall give rise to useful work." The extreme positivists are apt to restrict the possibilities of useful work. In the end, they return to the traditional sheep-or-goat concept of truth.

3. LITERARY CRITICISM IN THE WORLD OF EINSTEIN

THE literary critic can scarcely borrow the content or specific technique of physics for his own peculiar purposes. Although his "reality" is also a "creation," it is composed of a different kind of material and has to stand up against a different kind of pressure. He has to deal immediately with meanings that cannot be strictly verified or precisely formulated. He has constantly to form judgments and assert choices before all the operational returns come in, and can hardly even hope for an efficient method of getting such returns. To him even quantum physics is a model of law and order.

Yet he can profit by the example of modern physics. It is a striking demonstration of the general principles that in my belief should govern truth-making in all fields, and it enables the most effective statement of these principles. Physicists have revolutionized thought by the simple surgical operation of cutting the article "the" out of all key statements; Euclid's geometry is now *a* geometry, Newton's universe *a* universe, Aristotle's logic *a* logic. Nevertheless critics are still out for *the* genuine truth, and will accept no paltry substitute. Physicists announce calmly that very likely all their present theories will in time be proved wrong, and so they welcome new theories. "Failure to encourage fertility and flexibility in formation of hypotheses as frames of reference," Dewey writes, "is closer to a death warrant of a science than any other one thing." Nevertheless specialists in the good, the true, and the beautiful—or for that matter in the vulgar affairs of economics and politics—seem to think that to sign such a warrant is their first duty as responsible thinkers. Hence they lay down a barrage of dogma that permits all the breadth of outlook and ease of operation that one gets from a shell hole. And the more "scientific" they become, the farther they usually depart from the base of operations of physics. Orthodox Marxists and Freudians, for instance, are much closer to St. Thomas Aquinas than to Einstein.

In general, when the practitioners of the most exact science claim no more than an approximate, probable truth for their statements, all the dogmatists in the humanities must look simply silly. The literary critic too should employ a progressive, adventurous logic, especially because creative artists are as inveterate experimenters as scientists, and much harder to place or keep placed. He too should drag out into the open the fundamental assumptions implicated in his most innocent inquiry or simplest judgment, especially because unconscious assumptions are more dangerous in these more intimate concerns; like unconscious conflicts in neurotics, they become overloaded with emotion and therefore all the more immune to criticism. And once the critic has got his assumptions on the board, he can pick up from the physicist some general hints about how to play them.

Thus he should be as willing to put up with conflicting theories and tentative choices. One of the chief problems in physics today is that electrons appear to behave both like waves and like bullets; but although physicists hope to find a formula that will reconcile or comprehend the wave and corpuscular theories of light, meanwhile they use either according to the type of phenomena they are dealing with. So might the critic be content to take writers, who behave in all sorts of ways that no one formula can comprehend. Similarly he too might waive the ultimate, unanswerable questions—such questions as whether art is a manifestation of an immortal reality. Undistracted by the beyond-which-nothing, he could concentrate more efficiently on the business of correlating and evaluating esthetic experiences on this earth. Above all, he could test the efficacy of his principles. Scientists carefully work out and test the logical consequences of a theory, thus determining its usefulness, and they drop one that has ceased to be fertile even though it is not directly contradicted by experience. Critics are forever wrangling about the intrinsic truth of a theory, which cannot be absolutely proved; and though they cannot exactly determine its consequences either, often they do not even consider them. At least their thinking might be more fruitful if they concentrated on the pragmatic question of what *difference* a principle will make. This test exposes, I believe, the positive harm done by some critical doctrines, in cramping or distorting our experience in art. It also makes clear that some famous definitions of Art and Beauty make little practical difference, amounting to little more than the statement that art is wonderful. They give us a glow, which is something; but they are useless for further analysis or the actual practice of criticism.*

Far more significant, however, is the linkage between the arts and the sciences that is implicit in a fairly recent inquiry. Physicists and chemists are now doing considerable work on crystal structure and the pattern properties of molecules. Some seventy-one different compounds, for example, have the same molecular formula $C_6H_{12}O_6$; carbon atoms, which are regarded as identical individuals, form charcoal when arranged in one way, diamonds when arranged in another. Hence we get the concept of a whole that is different from the sum of its parts, a continuous structure that cannot be adequately understood by the ordinary devices of piecemeal analysis and mathematical measurement.

> We appear to be unable to grasp some of the relations of a whole [confesses E. W. Hobson] without breaking it up, as it were atomistically, and then proceeding to reconstruct the whole by a synthetic process . . . which by its very nature is such that the whole is never actually reached within the process.

This concept obviously applies to the work of art; it will concern us much more when we come to the other sciences. But meanwhile it points to the chief pertinence of modern physics for the literary critic. The recognition of pattern properties is but one example of the breakdown of rigid mechanistic concepts, of the assumptions and claims that produced a devastating split in thought. The all-important implication of the physicist's new view of the world is that it makes possible a reconciliation of philosophy, art, and science after long years of invidious specialization in different orders of "knowledge" and exasperated dispute over different levels of "truth."

* The word games of traditional esthetics are an obvious case in point. "Is a chair finely made tragic or comic?" asks Stephen Dedalus-Joyce. "Is the bust of Sir Philip Crampton lyrical, epical, or dramatic? If not, why?" And if so what of it?

This is again a familiar story; but again men have still really to get its point. Locke laid out the field of investigation for physics by isolating the "objective" qualities of matter (weight, size, shape) and shutting out the "subjective" qualities (color, smell, feel). He also made this distinction invidious by calling the former "primary," the latter "secondary." Thus Newton wrote that to speak of light as colored is to speak "grossly," in the manner of "vulgar people." Precisely like the theologians, then, physicists regarded their arbitrary logical scheme as absolute, unvarnished truth. Unlike the theologians, however, they made their scheme work; they could and did move mountains. While Berkeley, Hume, Kant, and other philosophers puzzled and haggled over the epistemology of science, or shattered the foundations of its concepts, laboratory men went on manipulating these concepts with extraordinary practical success, indifferent to all this metaphysical nonsense. "The scientists had found a metaphysics which suited their purposes," writes Michael Roberts, "and therefore proposed that metaphysical speculation be stopped, much as conservatives sometimes plead for the abolition of party politics." And so they created the mechanistic universe, as orderly as a formal garden, that so dazzled the eighteenth century. "God said, 'Let Newton be!' and all was light."

Yet this light threw long shadows—especially on the Soul. Where did the Soul come in? The remarkable efficiency of the methods for dealing with the objective, quantitative, measurable aspects of experience naturally devaluated the subjective, qualitative, and immeasurable. Physics came to be regarded as the most fundamental of the sciences, the supreme court that determined the constitutionality of all knowledge. Hence, though the great scientists were usually humble or even religious men, there developed the dogmatic materialism that denied reality to the inner world of feeling and steadily undermined the claims of art, religion, and all other philosophy. As a result, science was saddled with a rigid philosophy whose metaphysical assumptions were regarded as physical facts; art was driven into a corner where it made ineffectual claims to some superior brand of truth; philosophy and religion acquired a still greater contempt of the "material," a still more supercilious disregard of the importance and worth of the natural environment. Thought in general was blighted by the contradiction between accepted dogma and living consciousness—between the intellectual conviction that human behavior is absolutely determined and the intuitive conviction that man is a self-determining organism. "This radical inconsistency at the basis of modern thought," observes Whitehead, "accounts for much that is half-hearted and wavering in our civilization."

The blight is plain enough in nineteenth-century poetry. Wordsworth, who began by welcoming science, grew more and more peevish about "that false secondary power" that restricts us to a view of objects "in disconnection dull and spiritless"; Arnold listened miserably to the "melancholy, long, withdrawing roar" of the ebbing sea of faith; Tennyson agonized over the stars that "blindly run." Everywhere the specialists in spirit sought refuge in some cloudy, self-hypnotic idealism, a truth that seemed higher only because it had no visible foundation. And today literary men still strut or fret in the shadows of classical physics. On the one hand a simple worship of science, or an envy of its triumphs, has led to an indiscriminate use of its methods and materials, a substitution of information mongering for imaginative creation and esthetic analysis. On the other hand as simple a disillusionment, or a fear of its triumphs, has led to a scorn of the "naturalistic," a retreat to some elegant tradition or to private sensibility, a worship of the Blood or the Unconscious—ultimately to the whole flight from reason or holy war against it.

In short, the philosophy of modern physics has yet to become actually diffused in the climate of opinion. For that matter, it has yet to affect most laboratory workers. Nevertheless the truth remains that on its forefront science has itself criticized and supplemented the rationale that did such violence to man's concrete experience. It has itself made possible a more inclusive view of man and his world in which no vital interest need be sacrificed and all knowledge may become really usable. There are important differences in kind in the operations of science, philosophy, and imaginative literature, but the differences have their roots in important similarities. All alike construct an imaginative scheme, a kind of art form, the first vision of which is intuitive. The utterances of all contain an arbitrary, even irrational element, and are ultimately metaphorical. The utterances of none are final and absolute.

Again, however, warnings are in order. There is now an inclination to seek immediate spiritual dividends from an investment in science, to look for an absolute guarantee of values that is to be found nowhere under the sun. J. W. N. Sullivan, for instance, took off to the "perfectly possible" outcome of this reorganization of knowledge: "values will be established as inherent in reality." But science does not and cannot identify values with any essential, ultimate reality, make them inhere in anything but human experience on this earth. It has simply removed the grounds for disparaging them as subjective and therefore illusory, less real than the visible world of matter or the model world of atoms. It has recognized that all qualities are ultimately in the same boat; the "secondary" are not purely subjective, the "primary" are not purely objective, for both involve a human observer. It has dissolved the world of sense and common-sense perceptions but also restored its validity within limits—the world of physics too is one of appearances. It has made room for different modes of experience—the perceptions of the plain man, the artist, and the mystic as well as the scientist—by denying any one a claim to absolute objectivity. It has allowed us more freely to adjust these varied perceptions and to judge them on their merits, their practical usefulness in the common cause of more satisfactory dealings with a reality that in the end remains unknown.

In sum, what is implied in the philosophy of modern physics is no specific form of idealism or system of values but an organic, dynamic, functional, pragmatic view in which values and ideals may be shaped more realistically. And this in turn is essentially a restatement of the old humanistic tradition, as it has passed from Montaigne to Thomas Mann. In the philosophy of physics and humanism alike, the specific doctrinal content is variable, subject to revision in the light of our new knowledge or of changed conditions. Most important is the guiding attitude, the principles that generate the doctrine at any given stage.

4. SEMANTICS

As I have remarked, science is much more than "organized common sense." The great problem today is precisely to get it into common sense, the habitual attitudes of thinkers outside the laboratory. And the chief attack on this problem has been through semantics, the study of "the meaning of meaning." Perhaps the most significant of recent rediscoveries is that in the beginning of all confusion is, indeed, the Word.

I say "rediscovery" advisedly: the popular semantic reformers, such as Stuart Chase, often proclaim the hoary and obvious in terms of the novel and scandalous. Hardly a philosopher since Socrates has not remarked that language can confuse as well as disguise thought; plain men have always echoed Hamlet's "Words, words, words." Yet semantics is in fact a new study, and despite its enthusiastic disciples a profoundly important one. Today men not only are more acutely conscious of the havoc wrought by the ancient habit of mistaking words for things; they have a more acute idea of the nature of words. They are conducting a systematic inquiry into language: the relation of its symbols to one another, to the thing they symbolize, and to the agent who interprets this relation. They are developing a science of symbols. And here is where physics comes in.

To begin at the beginning again, there is no disputing the faultiness of our whole system of communication: the many blockages and gaps, the crazy tangle of cross wires, and all the careless, excitable, stupid, or ignorant operators at the switchboard. Thinkers have made a profession of using the deficiencies of language as the chief means to knowledge of Reality; their very logic has only systematized the confusions of common men. Nor is there any disputing the grave consequences of such habits, for the individual as for society. As Alfred Korzybski observes, we all live, are happy and unhappy, by what amounts to a set of definitions; and trouble begins when experience does not accord with our definitions. The "practical" man, who knows what he means and owns his world, may have the worst trouble, for his unquestioned common-sense definitions embalm the primitive attitudes implicit in the very structure of language. What he means is a myth, what he owns is a delusion. In general, serious discussion usually begins as an exchange of verbal gold bricks, and so ends as a hail of brickbats.

The most comprehensive, ambitious treatment of this whole problem so far is Korzybski's *Science and Sanity*. Carnap, Russell, Morris, and others are more brilliant or acute analysts, but they have been concerned primarily with syntax, epistemology, symbolic logic—the formal philosophical problems raised by semantics. Korzybski has tried to universalize the new orientation in physics and apply it directly to social behavior. For literary critics, at least, his work is a more useful text.

It might seem presumptuous to offer a brief summary of a work that the author insists must be read "diligently and repeatedly"—*"at least twice"*; but the core of it can be reached quickly enough. Korzybski has constructed a "non-Aristotelian system of general semantics" corresponding to the non-Euclidean and non-Newtonian systems in mathematics and physics. It is based on the assumption that all our knowledge is of order, structure, relations, not of substance or essence; it emphasizes especially the principles of "non-elementalism" and "non-identity." Our ancient language, with its substantive nouns, its *is* of identity, its subject-predicate propositions, its qualitative adjectives that are hypostatized as "properties" and finally as essences (as "conscious" is made into a noun, which then becomes a thing, an entity, finally a Soul)—this language, Korzybski maintains, grossly misrepresents the actual processes of knowing and the actual goings-on in the world.* In other words, it is as misleading as a false map. Men mistake its structure for the structure of the world, much as a child thinks of Great Britain as necessarily pink.

These linguistic habits, which are usually regarded as deplorable but after all natural and "human," Korzybski insists are literally unnatural, inhuman. Here he goes to biology, physiology, and psychiatry. His principle of non-elementalism corresponds to the biological organism as a whole, and more specifically to the structure of the nervous system; his denial of the *is* of identity then leads to his cardinal doctrine that man *is not* an animal. The animal's powers of abstraction are sharply limited, its reactions becoming unconditional at some definite point. Man's behavior is fundamentally different in kind because he can abstract indefinitely and is potentially capable of unlimited conditioning. Accordingly, when he arbitrarily stops abstracting at some point and calls it Reality, when he automatically responds

* Another way of describing the revolution in physics is to say that key nouns have been changed into verbs—to move, to act, to happen. *What* moves and acts, physicists do not care; "matter" to them means "to matter," to make a difference. But our language is still geared to express "states of being" rather than processes. In this connection, also, the German language helps to explain German philosophy. The Germans have been especially prone to hypostatize their abstractions, identify the Rational and the Real, invent concepts comparable to frankfurterness and sauerkrautitude—for they capitalize all their nouns. And this may help to explain their present worship of the State.

to his symbols as Pavlov's dogs do to a tuning fork, then he is "copying animals." He is overworking the lower nerve centers of the brain, setting up "psychophysiological blockages" to the higher centers in the cortex, disorganizing his nervous system. Worse than arrested development, he is exhibiting downright pathology; the delusions of the mentally ill are only intensified forms of the identifications and projections common to supposedly healthy men. Hence the *average* man is not *normal*. The philosopher, contemplating eternal entities in his role of privy councilor to the Absolute, is not godlike but conspicuously subnormal, unsane. The uncompromising revolutionist, contemptuous of mere reformers, is himself only a tinkerer; he explodes a particular set of abstractions only to replace them with a new set as unconditional and tyrannical, and so far from attacking the root evil, he is a striking symptom of it.

Nevertheless Korzybski perceives that Bridgman's operational test cannot be applied strictly outside of science, and that mathematics, which is language "of the highest perfection," is also language "at its lowest development" because it is restricted to a very narrow field of meanings. He perceives that in a world of multiple particulars involved in multiple relations, a ceaseless traffic in which every event has endless ramifications, our inclusive statements will naturally have a wide range of possible meanings. He perceives, generally, that there can be no bureaucratic solution of our difficulties, no perfectly organized language in which every symbol will have its number, its exact office to fulfil, its time clock to punch. So he does not go whooping after all the abstractions that have no precise "referents" or good references. His solution, instead, is *"a full consciousness of abstracting"*—of the relative, conditional, variable meanings of the positive, unconditional, one-valued statements we are in practice forced to make.

Abstracting is the characteristic activity of the nervous system, as indeed of all living protoplasm. The so-called concrete object perceived by the senses is itself an abstraction from a whirl of billions of electrons in space-time ("We *see* what we see because we *miss* all the finer details"); the name we give the object is an abstraction of still fewer general characteristics; the statement we make about it is a further abstraction; and so we can go on indefinitely through ever higher orders. The higher we go, the farther we get from objective referents—and the closer we come to a complete organization of knowledge, a complete realization of the potentialities of the human nervous system. But the higher we go, the more care we must take to distinguish between the different orders of abstraction—the scientific objects, the natural objects, the labels, the statements and the statements about statements, the inferences and the inferences from inferences; men who cry that Whirl is king are ascribing the shiftiness and impermanence of sensations to the higher orders of abstraction. We must be aware of the "multiordinality" or many meanings of our terms, so that we may properly limit, qualify, specify the context; "determinism," again, has caused so much anguish because it has been conceived as a one-valued instead of a many-valued term. And at all times we must keep in mind the multiplicity and uniqueness on the unspeakable level of immediate experience, the "is not" of our abstractions on all levels, so that we may never fall into the impossible assumption of "allness" in our knowings, never permit the convenient "is" to become the despotic "is nothing but."

Finally, Korzybski conceives his system as a practical educational enterprise, and has founded the Institute of General Semantics to carry it out. He has built a Structural Differential to represent the processes of abstracting, make visible that this *is not* that, and he is giving people workouts on it. It is a kind of machine for exercising the linguistic muscles, training all the nerve centers, giving the organism-as-a-whole the feel of the new concepts. He is confident that his training will enable "the average man or scientist to become a 'genius.' " As adults are hard to recondition, he admits that "it will take a generation" to root out the vicious linguistic habits; but children can be caught young and easily made into little non-Aristotelians. "The future generations, of course, will have no difficulty whatsoever in establishing the healthy semantic reactions." And enthusiastic converts write in, "We have found the way to world unity and peace."

Here I am left behind. Korzybski has the traditional faith in the philosopher's stone—the noble but somewhat unsane faith that by some principle of truth men can be conditioned to perfectly rational behavior. In theory he recognizes the importance of the products of the lower brain centers, known (loosely) as instincts, feelings, emotions; but in effect he treats them as poor relations who must of course be taken care of by the cortex but who will of course then behave themselves. His faith finally betrays serious omissions and inconsistencies in his system itself.

Throughout *Science and Sanity* Korzybski makes sweeping, positive assertions that are a little unbecoming to one who insists on a "general principle of uncertainty in all statements," but are at any rate inconsistent with his new structural trinity: non-Euclidean, non-Newtonian, non-Aristotelian. Non-Euclidean geometries are simply more geometries and have not destroyed the validity of Euclid; non-Newtonian physics includes most of Newton. The non-Aristotelian system, however, is a flat rejection of most of Aristotle; Korzybski asserts that the principle of identity is "invariably false to facts," and he plainly assumes that his principle of non-identity is "true." Moreover, he pushes this principle so hard that he obscures the kind of identity implied by uniformities and invariant relations. Specifically, he takes so literally his principle that "man *is not* an animal" that he makes uniqueness virtually as pure and absolute as the old essences. He neglects the underlying similarities in structure and function which, practically, justify the statement that man is also an animal. He neglects such primary biological drives as sex, because of which men perforce copy animals, and all the nonrational purposes rooted in these natural continuities. His semantic paradise could be realized only by creatures who strictly and absolutely were not animals.

Hence he merely evades the real problem when he asserts that the habit of "identification" is purely pathological. It is indeed a dangerous habit—but it is not perverse. It expresses a deep natural desire, satisfies a fundamental need. Without a large measure of identification, there could be no allegiance, no ritual, no society, no devotion to truth itself. Korzybski's "absolute individuals" can live together and communicate only because they do identify themselves with common causes, pledge loyalty to common symbols; if they took none of their symbols literally they could point to things more accurately, but they would have no purpose in pointing. In short, like the positivists and the popularizers, Korzybski finally slights the necessity as well as the strength of our deplorable linguistic habits. The semantic purist becomes fundamentally as unrealistic as the word-drunk schoolmen. While sternly demanding that language always be related to a behaving world, he overlooks the actual conditions of this world, the concrete social operations that give meaning. Even Humpty Dumpty was more realistic when he said that words mean what we *say* they mean.

It is thus significant that in "speaking about speaking" for some eight hundred pages, Korzybski ignores one of man's most characteristic modes of speaking—imaginative literature. In his system, this might be considered a peculiarly violent semantic disturbance, systematically exploiting the primitive habit of personification and identification. It might also be considered a peculiarly valuable means of orientation: ordering the sensations of immediate experience, translating the higher abstractions of intellect back into the terms of emotion and intuition, rounding out a perfect semantic curve. I imagine that Korzybski would take the latter stand. Presumably "consciousness of abstracting" could take the curse off poetry, and once he remarks in passing that a few sentences of poetry may "convey more of lasting values than a whole volume of scientific analysis." Still, the point is that he does not specifically include it in his all-embracing non-Aristotelian system, does not indicate how it conveys all these values or what the values are. If he does get around to this study, he will have to revise some of his principles, qualify his faith that handling a Structural Differential and learning never to mean *is* can satisfy the essentials of interest and purpose.

Even from the most radical semantic reformers, to be sure, I have not yet heard a demand for legislation against the writer's art. They seem prepared to tolerate it, if only for old-time's sake, so long as it is rendered harmless by a realization of its trickery, or as it is confined to an incidental embellishment of accepted meanings. But the trouble is that they merely tolerate it, do not really take it into account. No science of symbols can be adequate that does not investigate the use—the highly effective use—of symbols in poetry, and in popular speech. On the face of it, the greater writers succeed wonderfully in communicating the qualities of experience, work miracles of representation and realization impossible with the symbols of science. The language of poetry is unquestionably the most precise we have for the kind of experience it points to. Similarly popular speech is at its best accurate, economical, and efficient. The statement that a person is hard-boiled or stuck-up, a yes man or a rubberneck, conveys aptly a meaning that is no less intelligible and important because it cannot be put to a strict operational test. All in all, as Kenneth Burke sums it up, the semantic ideal is to evolve a vocabulary that will give the name and address of every event in the universe. Science already has such a postal system for the events it is interested in, and a method for checking up on whether its references arrive. But the name and address do not exhaust the significance of an event. Although we need to locate John Doe in order to do business with him, the business will involve subtle, intricate meanings that cannot be postmarked—all the complexities of human intercourse that sometimes make even scientists a little inefficient in their private affairs, and even mathematicians a little queer.

Of late, these complications have indeed received more attention. Charles W. Morris, for example, has been investigating the "pragmatic dimension of meaning" and even the nature of esthetic signs (though he does not get down to a concrete analysis of poetry). But more typical are such performances as *An Anatomy of Literature* by Edward M. Maisel, a student of Rudolph Carnap. "This novel cost $2 and was written by Harold Bell Wright" is a meaningful statement, he declares. Such a statement, however, as "This novel by Wright is inferior to *War and Peace*" is strictly nonsensical; it cannot be verified. He accordingly begins by demonstrating that almost all literary criticism is nonsense, and he proceeds as easily by throwing overboard all the most important questions of esthetic meanings and values; none of them are meaningful. And so it becomes necessary to announce the elementary. As purposeful creatures, we simply cannot afford Maisel's kind of sense. We cannot find exact referents and operations for all our needs, cannot base all our attitudes on fact. There are not enough facts to go around. Meanwhile there are plenty of exigencies; and we have to assert preferences, make decisions, and act. Communal effort, values and ideals, the ends of civilization, the very meanings that hold a society together—all rest on ideas impossible to verify.

To repeat, a science of meaning must take into account the intricate matters of connotation and feeling-tone, metaphor and personification, rhythm and pattern—the conscious tricks and unconscious habits that may blur meaning but may also intensify it, and that certainly amplify and socialize it. But to return to Korzybski, I do not believe that his stand is as definitely "*anti*-poetic" as Burke has declared. His constant attack on "identification" is not a logically necessary consequence of his basic philosophy; it is simply a natural consequence of his occupation as a philosopher. It is another sign of the rationalistic excess that lands him in inconsistency but that may be discounted easily enough. At any rate, Korzybski's working principles of the multi-valence of abstractions and the need of full consciousness of abstracting enable practical dealings with both logical and poetic symbols; and in this consciousness one may still "identify" himself with what his symbols stand for, pledge allegiance to certain meanings, precisely as one may have strong faiths and still hang on to his critical faculties. Qualified and supplemented, Korzybski's system can embrace both semantic and poetic meanings, and break down the artificial opposition between them.

Hence I should dwell on what he allows for in theory but overlooks in practice, chiefly to support his own belief that this system should not lead to overrationalization and "take the joy out of life." Throughout, Korzybski admits the importance of the lower brain centers: the field of "impressions," "feelings," "intuitions," all shifting, vague, impermanent, and "unspeakable," that nevertheless is biologically the oldest and longest traveled, provides all the material for higher abstractions, and is of vital significance in our daily lives. But he fails to develop fully the implications of his physiology, which are especially important for the theory and practice of art. In these terms, artists order and clarify this field, define the impressions, in a sense speak the unspeakable. More important, they may really assimilate the higher abstractions, translate them back into the terms of immediate experience—"visualization," "emotion," "sentiment," "culture." In insisting that the effort of anti-intellectuals to eliminate the higher abstractions and their effects simply disorganizes the nervous system, Korzybski does not sufficiently consider the opposite kind of disorganization, the hypertrophy of the cortex. "Semantic blockages" are as real in the passage from higher to lower as from lower to higher abstractions; with their deficiency in "feeling" and their frequent contempt of emotional needs, the superintellectuals would seem to be as pathological as the anti-intellectuals. They may literally lack sense.

In the whole task of humanizing our knowledge, imaginative literature may be expected to play a leading role. The critic still speaks chiefly in abstract, conceptual terms (as I am speaking now); the poet and the novelist vitalize these terms, round out the process.* Nevertheless the critic may act as middleman, and *Science and Sanity* can equip him with a sounder grammar and vocabulary. Unlike the positivists and the totalitarian reformers, Korzybski points to a principle of flexibility, not a fixation, as the solution of our problems. He leaves room for the necessary criticism and supplement of his thought: it is consciously dated Science 1933. He aspires to the "maximum of conditionality" and takes full account of both the possibilities and the responsibilities implicit in man's unique structure, his unique activity as a "time-binder." Specifically, he recognizes that formal definitions are no solution, and instead of trying to fix meanings tries to hang on to all their manifestations. If it is hard to feel confident that he is ushering in a new era, he remains an excellent guide in this one. The modern world demands at once unusual suppleness and unusual toughness of thought: an ability to accommodate oneself to rapidly changing conditions without losing one's bearings or identity, an ability to make ever finer discriminations without losing sight of the whole. Korzybski points toward this union of flexibility and firmness, this nice balance between the value of differences and the value of unity, that can relieve the strain and confusion of reorientation.

* Here is a way of defining the achievement of Thomas Mann: he has not only embraced an immense range of thought but penetrated and transfused it, reclaimed it for esthetic purposes, brought it into relation with the oldest experience of the race. Here is also a way of defining the limitations of other writers who have tried as earnestly to translate the abstractions of psychology and sociology, especially of Freud and Marx: they have not so thoroughly assimilated and recovered this material, and so they are apt to hand back too literal translations, abstractions not bodied out but merely draped or disguised.

V

BIOLOGY

But I must needs confess, I do not find
The Motions of my mind
So purified as yet, but at the best
My Body claims in them an interest.

SIR JOHN SUCKLING.

1. BIOLOGY COMES OF AGE

SINCE Darwin, biology has been a word to conjure with; but writers have conjured chiefly a host of false analogies, illogical inferences, fantastic hopes and fears. The clue to much of this misconception is the sophomore's discovery that love is nothing but an urge to propagate the species. Thus Aldous Huxley used to be distressed by the vulgar origins and affinities of human values; Theodore Dreiser still has by heart the lesson he got from biochemistry, that all human behavior is only an "inexplicable but unimportant" result of "chemic compulsions"; innumerable others persist gloomily in a confusion of beginnings with ends, the basis with the essence of phenomena. (Such distress must have hit a new low with the recent announcement that mother love has been isolated as a hormone.) Similarly, however, much arrogant optimism has also been inspired by biology. As a supposed guarantor of Progress, the theory of evolution has become the investment trust for all easy faith. And biologists themselves often have wild dreams about what they may do for the race. "No doubt," admitted one after discussing the hypothetical possibilities of ectogenesis, "many of the participants will wish also to continue in the old method of reproduction." It seems a rather perfunctory concession to the habits of some million years.

Such excesses, however, are a natural outcome of the peculiar difficulties confronting the biologist. His subject matter is distinguished by its immense variety and complexity, it involves many more specialized problems than does physics, and it is much less satisfactory for mathematical measurement and controlled experiment. Above all, it is intimately tied up with vital interests. Biologists are often harassed by outraged laymen, as in the controversy over evolution; whereas the nebular hypothesis, a theory of the evolution of matter, could be taught even in theological seminaries or the public schools of Tennessee. But they are themselves naturally affected by these same interests, and have themselves generated a great deal of waste motion and emotion. Thus their thinking was long distracted by such antitheses as heredity versus environment, structure versus function, teleology versus causation—antitheses that do not exist in nature but only in our ways of describing nature, and that as subjects of debate are about as pointless as the question of which came first, the chicken or the egg. And behind all such purely verbal issues was the flat opposition between vitalists and mechanists. The vitalists insisted that some altogether new principle—an entelechy, an *élan vital*—was necessary to explain life; the mechanists insisted that the principles of physics were not only adequate but essential. Both tended to lose sight of the living organism in their logical dispute over explanation. Both could have profited by the common sense of William Hunter in the eighteenth century: "Some physiologists will have it that the stomach is a mill, others that it is a fermenting vat, others again that it is a stewpan; but, in my view of the matter, it is neither a mill, a fermenting vat nor a stewpan but a stomach, gentlemen, a stomach."

Most biologists in the past were mechanists. They were naturally disposed to borrow the impressive apparatus of concepts and methods that physics had developed by the time their own science got under way. Moreover there were sound objections to the principle of vitalism: it was vague, uneconomical, unproductive for further inquiry, apt to let in again the unprofitable holy ghosts of metaphysics. Nevertheless there were also serious objections to their own assumption, even aside from their treatment of it as a plain fact. ("It is self-evident," wrote Max Verworn in the eleventh edition of the *Encyclopedia Britannica,* ". . . that only such laws as govern the material world will be found governing vital phenomena—the laws, i. e., . . . of mechanics.") The mechanists failed to account adequately for the most distinctive characteristics of living things—organization, development, growth. They had increasing difficulty in getting the established facts to fall into place in the accepted patterns. And then physicists began their drastic revision of the basic concepts of mechanics. Biologists were left standing firmly on a foundation that no longer existed.

Hence biology has recently undergone its own revolution. Mechanism was "self-evident" only on the basis of what Jennings calls "the fallacy of drawing negative conclusions from positive observations": by showing that some factor is operating, you prove that no other is. Workaday laboratory men, to be sure, still analyze isolate systems, treat protoplasm as a physical stuff and the cell as a lump-unit like the old atom; and their practice, within limits, is wholly legitimate. But the leaders in the field are now clearly marking these limits. (Jennings, Coghill, Child, Wheeler, Cannon, Herrick, Woodger, and Julian Huxley are among those I have in mind.) They assert that the traditional descriptions of biology have been not only incomplete but essentially inadequate. For the fundamental fact in biology, the necessary point of departure, is the organism. The cell is a chemical compound but more significantly a type of biological organization; the whole organism is not a mere aggregate but an architecture; the vital functions of growth, adaptation, reproduction —the final function of death—are not merely cellular but organic phenomena. Although parts and processes may be isolated for analytical purposes, they cannot be understood without reference to the dynamic, unified whole that is more than their sum. To say, for example, that a man is made up of certain chemical elements is a satisfactory description only for those who intend to use him as a fertilizer.

A good illustration of these "organismic" concepts in practice is G. E. Coghill's *Anatomy and the Problem of Behavior,* based on his experiments with Amblystoma. He discovered that the embryo of this lizard-like amphibian is perfectly integrated even before it has a nervous system. The basic factor is a potentiality of growth, a "forward reference" accompanied by a constant concern for the individual's integrity. As the total pattern expands, local patterns develop within it by a process of individuation; legs and gills, whose movements are at first inseparable from those of the whole trunk, acquire a measure of autonomy, become legs and gills in their own right. The nervous system itself emerges as a dynamic system whose role of conduction seems accessory; for a long time after it has definitely assumed this role it continues to grow, to develop a potential beyond the organism's capacity for expression in behavior—to "seek out new realms to conquer." It is as if from the beginning the creature conceives its end, carefully laying down a general plan of operations and an efficient store of reserves. The higher we go in the animal scale, the more pronounced is the "neural overgrowth" or "forward reference" that in adult man emerges as conscious foresight. In all animals alike, however, early behavior is not the mechanical cause but the result of growth, of functions implicit in the total pattern; the organism has a "creative potential" beyond the sum of its automatic reflexes.

Now at first glance this creative potential may look much like the old *élan vital,* and this account of little creatures bright with purpose may seem to open the door to all the old arguments from design and for the Designer. Actually, however, the organismic view is a higher synthesis of the vitalism-mechanism controversy, retaining the positive findings of both schools, transcending their artificial problems and their unnecessary exclusions. Biologists have indeed introduced a new principle; but it is a formulation of actual experimental findings and a basis for further experiment. It has already proved fruitful, as in the investigation of "organization centers" in embryonic development. Moreover, this principle by no means denies the physical basis of life or the basic continuity of the natural world. Organisms have a specific chemical constitution, none has been known definitely to violate physical laws, and all the specific findings of the mechanists remain valid. Organization implies mechanism, in fact, and the analysis of mechanism—as in anatomy and biochemistry—remains one of the important jobs of biology. The point is simply that physical and chemical laws, which are formulated to describe only the behavior of inorganic matter, do not describe adequately the much more complicated behavior of living organisms, and that an analysis of parts is apt to be seriously misleading unless it is supplemented by a study of wholes.* Vital processes are integrated with simpler inorganic processes

* In medicine, for example, the mechanistic concepts have done considerable harm, notably in the treatment of mental diseases and the prolonged hostility to psychoanalysis. The new attitude is summed up by C. W. Campbell: The physician "can no longer look upon the patient as the causal container of an interesting chemical situation, nor the chance battlefield of an obscure bacteriological combat"; he must consider the whole person in relation to the immediate and the larger environment, realizing that the misbehavior of stomach or bowel may be the result primarily of psychic disorders.

but are not identical with them. Life is not merely an added "property" of matter but something that transfuses and transforms it.

A new basis of operations always seems arbitrary at first. (Thus Leibnitz accused Newton of introducing "occult qualities and miracles into philosophy.") Hence the question is commonly asked, just what *is* this organic or creative power? But the question is itself the product of a language geared to ancient assumptions. If one begins with stuff, there is always the problem of getting the stuff to grow, just as in physics there is the problem of getting the little lumps to push one another around. But if one begins with process, dynamic organization—as physicists now do with energy—the problem vanishes. Nor is this to make it vanish by waving a verbal wand. To take stuff as the basic reality is at least as arbitrary; what is immediately given in experience is not states but events, and as for reflection, natural science knows nothing static in the whole universe. Organization and growth are at any rate the given data in biology, not logical inventions or deductions. As J. H. Woodger points out in his *Biological Principles,* the concept of the organism-as-a-whole leads one to expect what biologists once felt was a problem forced on them. There is no antithesis, for example, between teleology and causation. "Purpose" is not imported into nature, and need not be puzzled over as a strange or divine something else that gets inside and makes life go; it is no more an added force than mind is something in addition to brain. It is simply implicit in the fact of organization, and it is to be studied rather than admired or "explained."

Admittedly, these new concepts are still pretty fuzzy. They have yet to be formulated adequately and provided with a complete methodological apparatus; biology has yet to have its Newton, not to say its Einstein. Almost certainly it will never be able to compete with physics in exactness. Life appears to lack the lovely mathematical structure of the inorganic world; the organism-as-a-whole, which is more than the sum of its parts, defies arithmetic, and its processes cannot as yet be expressed satisfactorily in equations. (The Great Mathematician seems to have let down his creatures in the realm of their most vital interests.) Meanwhile, however, the organic concepts meet the essential pragmatic test of science: they embrace more facts, they make possible more fruitful work. It is important that the peculiar problems of the science of living organisms are no longer being evaded for the sake of a factitious exactness. As Woodger says, "Biology is being forced in spite of itself to become biological."

2. THE EVOLUTION OF EVOLUTIONISM

THE great generalization in biology, and one of the great ideas of the nineteenth century, remains the theory of evolution. Although the idea is at least as old as Anaximander, appearing even in St. Augustine, before Darwin it had been only a myth or a speculation. Darwin made it a working scientific hypothesis, supported it with an impressive mass of evidence, and in his theory of natural selection provided it with a reasonable *modus operandi*. Difficult questions remain; but some kind of evolution is now universally accepted by biologists, and some kind of evolutionary theory has entered all other fields of thought. Aside from the religious fundamentalists, moreover, men are no longer outraged by the idea. Most thinkers find it intoxicating and ennobling. They feel as Darwin did:

> There is grandeur in this view of life, with its several powers, having been originally breathed by the Creator into a few forms or into one; and that, whilst this planet has gone cycling on according to the fixed laws of gravity, from so simple a beginning endless forms most beautiful and most wonderful have been, and are being evolved.

Indeed, the idea has been too intoxicating; and so we had best come at once to the problems it has raised. These may be gathered under three main headings: the cause, the method, and the implications or consequences of the evolutionary process. The first question, *why* evolution should occur, is again unanswerable—outside of religion—but again less troublesome if process rather than substance is postulated as the fundamental fact; granted process, men will naturally expect to find some order in it. Nevertheless the mystery warrants some humility: it is strange, if mere survival is the end, that living organisms should ever have evolved to embark on their hopeless competition with the everlasting rocks. For that matter, considerable mystery still surrounds the method, the question of *how* evolution occurs. Though biologists have discovered a great deal since Darwin's time, especially about the mechanism of heredity, the genes, the pattern is far from complete; at best, their account of random variations and the processes of natural selection is less satisfying than the classical demonstrations of physics. In any event, this is a problem for the specialists. But the third question, of the ends and implications of evolution, is a necessary concern of all thinkers.

It has accordingly inspired the usual simplicities. Still popular is the notion that evolution moves in one line and one direction, invariably from "lower" to "higher," inevitably toward perfection. Actually, we have a branching development: both birds and mammals evolved from a reptilian ancestor, and insects come at the end of a separate long line. Even this development, moreover, is not so orderly as biologists would prefer to have it. There are apparent exceptions in the general trend from simple to complex, as in the atrophy of such complex organs as the appendix, or in the existence of parasites who are demonstrably simpler than their ancestors; there are apparent regressions, as in the stooped posture and clumsy gait of the Neanderthal race some 400,000 years after *Pithecanthropus erectus* had learned to walk like a man. Worse yet, nature chooses to select and encourage some variations that in the long run seem useless, even harmful. One example is a sexual display character like the train of peacocks or the antlers of certain deer; today it is easy enough to find analogies in human society. Aside from such anomalies, at any rate, there is no warrant whatever for the assumption of the inevitable progress, or even the long continued existence, of the human race. The simplest forms of life, such as the amoeba, continue to flourish while complex species come and go; certain insect societies antedate man by countless millennia and may well outlive him by countless more. And the whole spectacle, which in one aspect is so edifying, in another is a frightful exhibit of waste, stupidity, and cruelty. God has reason to resent His election as the author of such a show.*

Biologists themselves have oversimplified evolutionary theory in other ways. They did not immediately welcome it—"Systema-

* There is also some evidence, in a very long view, for universal *devolution*. A current theory in physics, based on the second law of thermodynamics, states that the universe is steadily running down, tending toward "maximum disorganization," blazing away to "heat death." In this view, the building-up processes on our earth are only a local, accidental, irrelevant interlude. For that matter, devolution is the probable fate of any one species or subspecies; for most have degenerated and finally become extinct.

tists," Darwin noted, "are far from being pleased at finding variability in important characters." When they did accept it, they fitted it into the neat, rigid scheme of mechanical physics: evolution was an automatic unfolding or *evolutio* of predetermined forms, just as the oak is implicit in the acorn. The clockwork universe was simply telling time. Hence they missed the most revolutionary implications of Darwin's theory, which undermined this Aristotelian scheme of fixed objects, fixed categories, and fixed ends. It emphasized process rather than the forms the life process happened to take, and "natural selection" suggests that these forms were not given in advance. Later investigation tended to confirm the random variations that Darwin suspected, making the theory of preformation still shakier. Meanwhile Samuel Butler and others were insisting that the creative energy of man must be considered as well as natural selection or the force of environment. And finally the facts before them compelled biologists to put new questions to nature.

Hence there developed the various theories of "emergent" or "creative" evolution, fathered by C. Lloyd Morgan and Samuel Alexander. These are based on the organismic concept of a whole different from the sum of its parts; the gist of them is that a new quality of existence "emerges" from combinations, a quality that is nonadditive and nonpredictable from a knowledge of the original elements. The favorite analogy is the emergence of water: water is qualitatively different from both oxygen and hydrogen, and no chemist could have foretold its quality. Just so, on the large scale, did life emerge from matter, mind from life. Once emerged, then, the new elements react upon matter to create still other combinations with new properties; living organisms are particularly active opportunists. For this principle and its results there are various other names: "holism," "collective novelty," "creative synthesis," "creative resultants," "epigenesis," etc. The theories differ in detail, especially concerning the main "levels" attained or implied by the evolutionary process. But their essential import is the same. The universe is not a machine grinding out mechanical wonders on a rigorous schedule, but a dynamic system that produces new elements and therefore new laws. Reality itself is on the make, things do rise above their source, there *are* new things under the sun.

This theory may also look like vitalism in a new disguise. Scientists have criticized it as a purely verbal solution, an elaborate confession of ignorance; they say that to call an effect an emergent is to give up all effort to explain it. A. O. Lovejoy answers, however, that they are demanding a special *type* of causal explanation—mechanical preformation. The theory does not deny the fundamental uniformities of nature or the kind of determinism—cause broadly conceived as regular sequence—that the scientist needs for his working purposes. Emergence is not represented as the effect of a metaphysical agency. It is due to a specific property of matter, an inherent cause, and it is continuous from the inorganic through the organic realm. Each level rests upon the lower—life upon matter, mind upon life—and is not a complete novelty but a realization of potentialities.

Nevertheless Jennings seems a little overoptimistic when he calls the theory of emergent evolution biology's Declaration of Independence. At least, biologists have yet to learn what to do with their freedom. They are unable to analyze this new property of matter or to isolate this new cause; in the laboratory they continue to treat their materials as resultants rather than emergents, to look for invariants rather than unpredictables. As now formulated, the theory still belongs in the realm of philosophy. (It was anticipated, in fact, by Schopenhauer.) It is not a clear-cut working hypothesis for scientists, it cannot be proved. Conceivably, moreover, it may be disproved: further knowledge may enable us to predict the emergents and may destroy their apparent novelty. Meanwhile the theory has its dangers. In their preoccupation with the part-whole relation, its devotees tend to neglect the external relations with the environment, the resultants that can be known. They also tend to hypostatize their new names, as when J. C. Smuts erects wholes into the sovereign principle of Holism, and to grow too ecstatic over their mysterious emergents; they run to capital letters—Life! Mind! maybe Deity! They tend, in short, to dream in their philosophy of more things than are found in heaven or earth.

These excesses are most apparent in the debate over the "levels" of emergence. The term itself is an unfortunate metaphor; it implies something static and spatial, obscures the dynamic process and organic unity that the creative evolutionists wish to stress. It has at any rate encouraged them to become too schematic. Granted continuity, the emergents are relative and the levels more or less arbitrary; but the tendency is to make too absolute distinctions and draw up too rigid hierarchies. Morgan was content with the three obvious levels of matter, life, and mind; Smuts added a fourth, Personality; Alexander had seven; G. P. Conger made out as many as twenty-five. Plainly, men are drawing these lines, not nature itself. Once they have started up the scale, moreover, it is hard for them to stop. Smuts calls a halt before he reaches the Wholiest; he declares that the next level must really be unpredictable. Alexander, however, goes all the way to deity; though this level does not yet exist, the whole process betokens it. (Thus God, too, is literally on the make.) This is a legitimate speculation, of course. But it is scarcely a fruitful one for biologists; by now science is completely out of sight. And by now we might well return to the common sense of W. M. Wheeler: "To the observer who contemplates the profuse and unabated emergence of idiots, morons, lunatics, criminals and parasites in our midst, Alexander's prospect of the emergence of deity is about as imminent as the Greek Kalends."

Yet it is easy to discount such extravagances. One can rub out the exclamation points, put the levels in the lower case, forget the mystery-mongering. One may even dispense with the unpredictability of emergents; all that is necessary to the theory, I should venture, is their qualitative uniqueness. There remains a coherent restatement of the theory of evolution, with a firm foundation in observable facts, and with a fuller recognition both of the life process itself and of its diverse products. Creative evolution is the kind of unifying principle or more inclusive pattern that science always seeks, for both pragmatic and esthetic reasons. For philosophy it is a pertinent restatement of monism. For biology it points to a way out of such problems as the preformation-epigenesis conflict in embryology by showing that there are both continuity and discontinuity, that evolution is both repetitive and creative. For thought in general, it is valuable for its recognition of unique qualities that cannot be treated merely in terms of quantity, its restatement of the peculiar problems of the science of life. If it is still too far in advance of the experimental data, and indeed too rich a hypothesis, it is nevertheless another approach to the new concepts that biology has needed, not merely for the sake of philosophical respectability, but for practical experimental purposes.

3. BIOLOGY AND PHYSICS: THE PATTERN OF MODERN SCIENCE

In the patriotic enthusiasm over biology's Declaration of Independence, there has naturally been considerable oratory. At times one gets the impression that biologists have cut all ties with the motherland of physics. In truth, they have not: their independence is largely nominal. Ideally, they should not: continuity remains a primary article of faith for all scientists. It is therefore desirable to return to the motherland and get another perspective on these goings-on.

Physical concepts are always as simple as physicists can make them. The atom is getting distressingly complicated, indeed, but physicists themselves are the most distressed; their chief hope and effort is to simplify it again, make it handier to work with. We should therefore not be surprised, Sullivan reminds us, if their atom is inadequate for the phenomena of life. Conceivably they could, if they set their minds to it, invent an atom that would be suitable for the purposes of biologists. And they may very well have to. Experimental science is constantly narrowing the gap between the "levels." "Life," for example, is difficult to define because no one quality that the biologist asserts for living organisms—irritability, capacity for reproduction, ability to effect repairs, etc.—cannot also be found in some form of inorganic matter. Theoretical science is working toward unifying concepts, with emphasis upon energy, process, the event rather than the object. Specifically, indeed, the trend in all the sciences is toward the organismic concept.

This whole development is so significant, so rich in implications and possible consequences, that there is no longer room for local patriotism. Biology and physics are more intimately related now than they actually were in the palmiest days of mechanism. One may point, in the first place, to certain general analogies. Both the random mutations and the emergents of current biological theory correspond roughly to the discontinuities and uncertainties implicit in quantum theory (though in neither, to repeat, is there a denial of fundamental *connectiveness*). Similarly biologists are employing something like the field concepts of physics and paying more attention to time. An organism is not an anatomy to be dissected in timeless space; it is an event with a field, a temporal-spatial whole that acts as well as reacts. Both sciences are now more like a functional physiology than a mere anatomy. Both emphasize the structure rather than the constitution of their materials, and structure viewed as a form of organization of energy. The cell, the molecule are not aggregates but integrates; the cell can do more things because it is more highly organized, not because it has some mysterious ingredient.

These linkages, moreover, are not the product of "mere" theory. They become more striking in the laboratory, the home of the hardheaded. In physics and chemistry, to repeat, investigation of crystal structure and pattern properties has made the total pattern or organism-as-a-whole a "reality" in physical science. In biology, on the other hand, C. M. Child has worked on "physiological gradients" in organisms: streams of energy radiating from a region of intense activity, passing to the less active regions of protoplasm, and indicated most conveniently by changes in electrical potential. Child can alter behavior and growth by altering this "rate of living" in different places in the organism—making a worm grow two heads or none at all, or one where its tail ought to be. And though such scientific tricks cannot be played on man, with his far more complex organism, he has become the subject of even more striking electrical demonstrations. The conduction of a nerve impulse, which may lead to a very spiritual idea, involves not only physiological work done—oxygen used, heat given off—but a measurable change in electrical potential; the recently discovered "brain waves" give some literal meaning to brainstorms. F. S. C. Northrop has therefore asked the critical question: "Is field physics applicable to living organisms? Are they electro-magnetic systems?" George Crile had already returned an emphatic yes in his "radio-electric interpretation" of life (*The Phenomena of Life*), but Northrop and Burr have accumulated more impressive evidence. They have measured potential differences within the body and discovered variations accompanying certain activities; the exact time of ovulation, for example, can be marked by a pointer-reading. They have also picked up potential differences on the organism as a whole, differences that remained remarkably constant. Individuals as well as species have their characteristic electrical patterns—presumably explaining why different men are affected differently, say, by sea air and mountain air. But we all seem to be surrounded by an electromagnetic field that is persistent in both time and space.*

Thus biologists again return to physics for basic concepts. It is significant, however, that in returning they carry with them their own concept of the organism and still recognize unique differences. If biology is not actually independent, it is no longer slavishly dependent. In fact, it is beginning to pay off its debt with suggestions for physics. Because their own methodology is still far better systematized, physicists are naturally reluctant to accept these favors. R. B. Lindsay, for instance, protests against the suggestion that they use "emergents" instead of their present descriptions. "The whole purpose of *inventing* atoms, molecules, nuclei, and electrons is to enable us to describe the large scale properties of matter"; it would be foolish to conceive these elements and then reject the properties they were invented to describe. Nevertheless physicists may be forced to take some such step. Their present inventions have landed them in considerable disorder: the electrons have become perverse, appearing to evade the solemn laws that the great stars are content to obey. Sullivan, himself a mathematician and physicist, accordingly declared that "the great intellectual conquests of the near future, the next great step forward in the development of the human consciousness, may be expected to come from biology." Even now, when no biological idea seems like a staggering intellectual achievement, especially to one bred on Einstein and Planck, the science of life points to a way. Sullivan agreed with Whitehead that the notion of particle will very probably have to be replaced by the notion of organism.

Meanwhile biology has one salutary lesson for the parent science. While physicists are now apt to get lost in the interstellar spaces or inside the atom, biologists are still close to the plain facts of existence on earth. Because they cannot express so much with equations, they keep their eyes on the object; with geologists, they accumulate evidence of the existence of a solid world long before there were men to read the pointer-readings and think the thought stuff; they deal with things like worms and frogs, babies and bellyaches, that can scarcely be considered mere

* If one shudders at the thought of what the imaginative will make of this idea, it nevertheless has practical bearings. Northrop and Burr have lately reported that incipient cancer can be discovered even before it can be seen or felt: because cancer cells deviate from the normal patterns of the "electrical architect," they disturb the normal electrical routine.

fictions. If only because their patterns and equations are less elegant, their theoretical accomplishments less astounding, they make clear the positive gains in positive knowledge.

4. BIOLOGY AND THE HUMANITIES

THIS account of modern biology, needless to add, is far from comprehensive. Specific findings in the various branches may well prove much more significant for the society of the future than any of these general ideas. Geneticists, for example, have made remarkable discoveries about the mechanisms and processes of heredity, opening up fantastic possibilities of man's controlling his evolution.* But my present aim is only to get at the main concepts, work out their broad implications; and these, I believe, are significant enough.

They also involve enough hazards. The diversity of material that gives the study of biology its cultural value also makes all generalization difficult and all analogy dangerous. The basic facts are that man is an animal and that he is different from every other animal; the critic must always maintain a delicate balance between these facts of biological continuity and biological uniqueness. The important thing, however, is that biologists themselves now insist on such cautions. Literary men tend to ignore either the continuity or the uniqueness; like Babbitt and Eliot they sharply oppose nature and human nature, like Zola and Dreiser they absolutely identify human nature with its physical or animal basis. Biologists answer that the life of man is the outcome of certain natural processes but by the same token is not identical with these processes. And they also answer such knowing critics as Michael Roberts and D. G. James, who begin with an excellent account of the revolution in science but end by missing its point, asserting that the imaginative scheme of

* Because these possibilities have become still more fabulous as they are popularized, some sober reminders are in order. In *Out of the Night,* Hermann J. Muller points out that centuries would be required to take even the first step, of eliminating feeblemindedness. As this is a recessive trait, carried in the genes of many normal people, the sterilization of all the feebleminded would reduce their number only by half in eight generations. Muller also points to a more serious complication in producing the scientifically well-bred man: society will first have to decide what is the superior type, what traits should be bred in and bred out. Many of us would not care to entrust this decision to the present rulers of society.

science is necessarily mechanistic, that it cannot admit "purpose," and that it is therefore irrelevant to the values associated with conscious purpose.

For the specific purposes of literary criticism, in the first place, the organismic concept offers obvious leads. A whole esthetic can be written in its terms—as we shall see when we come to Gestalt psychology, which systematically applies it to our direct experience. But the most important implications for criticism are the implications for the humanities in general. If "Know thyself" is the first principle of wisdom, biology is the logical place to begin any study of values. Values of every kind are finally notions about what behavior is appropriate and what experience good. These notions in turn must finally rest upon some conception of "human nature," its essential needs and bents, its given limitations and possibilities. Yet thinkers shirk the whole problem by writing out blank checks against human nature, and then filling in whatever amount is needed to purchase their preferences. Theologians have justified their profession of salvation by representing man as an utterly miserable, vile, despicable creature whom it would seem pointless to try to save. (The *De Miseria Conditionis Humanae* of Pope Innocent III is the crowning example.) Romantics have then insisted on man's natural goodness, attributing all evil to the institutions of Society; though they failed to explain how the happy good men came to mold such institutions and why they display such a perverse preference for them. Above all, men have persistently clung to what John Stuart Mill noted as the most vulgar of ethical fallacies, that of "attributing the diversities of conduct and character to inherent natural differences." And debaters in distress still flee to the old havens. "You can't change human nature," they announce; and then there is no getting at them. They have formed an axis with Jehovah.

Even more troublesome today, however, are again the silent, vague, or self-contradictory assumptions. Conservatives who regard human nature as immutable, given for all time, are nevertheless all too conscious of radical change. Usually they regard it as change for the worse—a falling from grace, a heading for hell. But liberals are typically no less arbitrary or ambiguous. In talking of the "natural rights" of man, they too are apt to appeal to inalterable principles of human nature, as the easiest way of guaranteeing their values and condemning changes they do not like. In short, all thinkers perforce take some stand here, but few have thoroughly examined the grounds or consistently faced the consequences of their stand. Their values are pegged on a mist.

Now, aside from the present limits of his knowledge, the biologist is in no position to serve as the final arbiter in any given controversy. "Human nature" is an abstraction, and can never be isolated or examined directly; we know only specific types of men, always under specific conditions. The social sciences can accordingly give more light on this whole issue. Meanwhile, however, biology can give us some information about basic structure and functions, some general principles to guide us through the particular controversies. Thus it refutes the doctrine of infinite perfectibility: man has a definite structure, structure determines potentialities, potentialities imply limitations as well as possibilities. But it also refutes the more common and more dangerous doctrine of the immutability of human nature. It gives us the long evolutionary view, in which we can see the profound changes that have taken place since man dropped his tail and took to the ground. More important, it emphasizes the remarkable plasticity of his present structure, the remarkable adaptiveness that has not merely been superimposed upon instinctive behavior but has transformed all his behavior. Man alone of animals can look before and after, think of what is not, *promise* what is not. Because he can foresee the future, he has a future; because he can promise, he has promise.

In *The Wisdom of the Body* Walter B. Cannon describes the marvelously intricate, efficient organization by which the body regulates its oxygen supply, internal temperature, blood content, and the other vital materials and processes of the internal environment. By this automatic control of routine necessities man is given much more freedom in his dealings with the world. But freedom for what? Chiefly for the full development of consciousness, Cannon answers: "The functions of the brain which subserve intelligence and imagination, insight and manual skill, are set free for the use of these higher services." One could grow sentimental about the selfless fidelity of the body, so often maligned as the cause of frustration of high endeavor while a nameless "we" is credited with all triumphs. But it does not bear alone the white man's burden. The "higher services" are not luxuries; they are indispensable *services*. In a mechanistic scheme, ideas are left without any real function and seem less creative, in fact, than the instincts of worms. In a dualistic scheme, they are given reality, usually a most distinguished kind, but are left without any clear means of exercising their function. In the organismic scheme they are still far from being thoroughly "explained," but they have a local habitation as well as a name: a definite scientific standing, a definite work to do, a definite mechanism to work with.

Hence biology also points to a way out of the ugly dilemma of determinism, and also without supporting the loose notions of indeterminism. "To say that the arrival of a train in Berlin is undetermined," writes Einstein, "is to talk nonsense unless you say in regard to what it is undetermined. If it arrives at all it is determined by something." The behavior of organisms is plainly lawful, the business of science is still to determine how they arrive. But to get this important information biologists have felt obliged to modify the immutable necessities imported from physics and admit a principle of conscious purpose, self-control, "free will." Man has clearly been able, in a brief hour of biological time, to make over both himself and his environment, to create new necessities. He has made such marks on nature, God help him, as a dust bowl. And the dust bowl, an altogether natural, lawful result of human behavior, is not an inevitable outcome of human existence; by as natural, lawful means man can deliberately repair the damage. Within the necessities imposed by nature he exercises the power of conscious, voluntary choice. In physiological terms, this power may be described as an amplification of the "forward reference" Coghill found in the nervous system of Amblystoma. And in such terms Coghill sums up the case against mechanistic behaviorism:

> If, then, it is conceded that growth is one of the means by which the nervous system performs its function in behavior, it must be granted . . . that man is more than the sum of his reflexes, instincts and immediate reactions of all sorts. He is all these plus his creative potential for the future. Even the embryo of *Amblystoma* is, mechanistically considered, more than the sum of its reflexes or immediate behavior possibilities. The real measure of the individual, accordingly, whether lower animal or man, must include the element of growth as a creative power. Man is, indeed, a mechanism, but he is a mechanism which, within his limitations of life, sensitivity and growth, is creating and operating himself.

Still more important, biologists emphasize the responsibilities of this privilege. Rousseau and Carlyle regarded consciousness as a kind of disease, arguing that we are not conscious of our organs when they are acting perfectly; today many writers go further and regard intellect as a curse—it stunts natural instincts, it stifles "the voice of the blood," it is the great enemy of all that is deeply "human." And too often this dangerous attack on reason is opposed merely by a more fervent appeal to reason. There are good biological grounds for the attack, indeed, when intellect is cultivated as an end in itself instead of as an instrument of the organism-as-a-whole. Nevertheless biologists also tell us that control is essential to efficient behavior in any organism, that conscious intelligence is incomparably the finest instrument of control yet evolved, and that it is now not merely an added reserve or overdrive but part of man's necessity. Because he has the power of voluntary control, he lacks the animal's mechanism of instinctive control. A man who invests his trust in natural instincts cannot begin to compete with an insect.

Here we return to the implications of evolution. In them, I believe, we must find the ultimate authority for our values. But they also require a long view and a broad view, they are all too easily dragged out and dressed up to suit any occasion; and so once more we need to pause. The apparently edifying faith of the Victorians in automatic progress actually obstructed a realistic consideration of the means and ends of a better society; the faith became a gospel of reaction. On such grounds they opposed social reform and made the whole cosmos bear the brunt of a fiercely competitive morality. Today we have many subtler versions of the principle that whatever is, is biologically right; men argue as if human efforts to "interfere" with natural processes were not themselves natural—and had not, on the face of it, met with considerable success. "Nature," declared Aristotle, "makes nothing either imperfect or in vain." Evolution plainly suggests otherwise. But by implication, at least, men still make a deity of nature—a peculiarly ineffectual deity that is obliged to do whatever happens and recommend whatever they want to happen.

More difficult, however, are the issues raised by the notion of "efficiency" that I have been employing. In a sense, man is undoubtedly the most efficient animal, in that he is the most resourceful, can do more things about a given situation. But there is also no doubt that the things he chooses to do are often hopelessly inefficient. No other animal is so persistently incompetent as a fool, no other species is so often bent on self-destruction; in any event the amoeba is a good bet to survive man. Furthermore, it is hard to demonstrate that all his distinctive qualities—his esthetic sense, for example—are an aid to him in the struggle for existence, or that what he generally considers his highest achievements have always made for greater security. The glory that was Greece means Athens to us, but in a struggle we would still put our money on Alexander. Briefly, the whole notion of biological fitness is not so clear as it might be.

Yet neither is there a clear way of retracing evolution. One may envy the simple savage or the contented cat, but one cannot by taking thought return to their state; if the kind of destiny implicit in the structure of man's brain does not bring happiness, then happiness is not for him. The wants and needs that may seem fantastically unnecessary are now actually part of his necessity. The testimony of his entire history, moreover, is that they should be trusted, for in satisfying them he has also established himself triumphantly in the natural world. Through the luxury of scientific speculation, for example, he has achieved an amazing control of his environment. "Prune down his extravagance, sober him," wrote James, "and you undo him." Biological fitness is relative, biological perfection strictly meaningless; but the natural aim of man is still a complete understanding and control of the conditions of his existence, the natural ideal is still a complete realization of his distinctive potentialities. And so the broad pragmatic, naturalistic justification of art is not merely that an esthetic sense is a given part of his equipment, but that art illuminates the other conditions of his existence, intensifies his realization of what he was and is and may be, at once deepens and heightens his consciousness of what it means to be human.

At any rate, biology does not endorse the popular notions of the survival of the fittest by the law of the jungle—the attitudes that are bearing such appalling fruit today. The opposition to such attitudes is again likely to occupy lofty, abstract moral grounds; but the firmest ground is the biological fact of coöperation. Coöperation has made possible the development of a civilization, and it is now plainly essential to survival. From the beginning, however, it was rooted in animal instinct; and after years of emphasis upon the struggle for existence, biologists have recently begun to study group behavior, the coöperation that modifies this struggle and is also a force in evolution. In *The Social Life of Animals,* for instance, W. C. Allee masses considerable evidence of partially social behavior throughout the animal kingdom. Animals are rarely solitary but almost necessarily members of loosely organized communities, and the community has clear advantages. There seem to be occasional anomalies; small groups of gold-fish and cockroaches sometimes do better jobs in learning a maze than do larger groups. But in all experiments so far, group behavior has differed from the behavior of separate individuals.

Indeed, coöperation begins in simple physiology, since organization implies social correlates all the way down to the cell, and from these humble sources biologists have drawn some edifying conclusions. Thus Child suggests that the trend of evolution is toward democracy. The simpler organisms are rigid totalitarian states; the "physiological gradients" at the center give orders and are little affected by the outer regions they influence. In man, however, the brain is more like "a parliament or deliberative assembly," which gives orders only after it has weighed the various appeals coming from these regions. (Compare Nietzsche's assertion that "our organism is on an oligarchic model.")

> We may say then [Child concludes] that in the higher vertebrates the gradient pattern of organization is undergoing modification from the autocratic form of dominance or control characteristic of the simple gradient toward the democratic form with representative government, with all its plasticity, modifiability and uncertainty of result in any particular case.

Likewise Cannon draws the democratic moral from "the wisdom of the body." The marvelous flexibility of the body's system of controls suggests that social stability is not to be sought in a rigid system. Similarly the rights as well as the responsibilities of the individual have their physiological basis: the cell retains its intimate texture in all the changes in which it takes part, and all the organic processes involve considerable self-regulation. In short, "the integrity of the organism as a whole rests on the integrity of its individual elements, and the elements . . . are impotent and useless save as parts of the organized whole."

Now such analogies are merely suggestive and prove nothing. Nevertheless the suggestions are sounder than the arguments for a despotic state that have been drawn from the static, stereotyped ant societies, or from the order of "peck-right" in a flock of hens (by which each bird pecks all those below it in the hierarchy and is pecked by all those above). Child and Cannon take into account the whole trend of evolution, the essential differences between men and ants or hens. In any event there remain certain indisputable facts, notably the fact of organization. Coördination and coöperation, unity and community are implied in it; and it is of supreme importance for human society, the highest order of integration yet achieved by life on this planet. Hence the humane values, the golden rule, the sentiment of solidarity are not the absurd ideals ridiculed by the Paretos nor the slave morality despised by the Nietzsches. They are outgrowths of biological necessities. Here is also the sound objection to the Gospel of Force and the Gospel of Uniformity. Regimentation is not integration; to demand utter conformity is to defy the fact of unique individuals and to stifle the possibilities of organic growth. Even a coat, Mill observed, has to be made to a man's measure; and it is certainly no easier to fit him with a life.

The sum of all this is again an argument for the traditional values of humanism. Biologists are agreed that the broad stream of evolution, despite all the eddies and apart from all notions of "progress," has flowed in the direction of greater complexity, finer organization, closer coördination. In the words of J. A. Thomson, it has flowed from the organized purposiveness of the amoeba to instinctive purpose, to perceived purpose, to conceived purpose. Its farthest reach has been the consciousness of man, as intensified, enriched, and extended by a collective culture. The natural good, accordingly, is all that contributes to this end; one makes the best of life, as Malraux's spokesman declares in *Man's Hope,* "by converting as wide a range of experience as possible into conscious thought." The natural evil is all that blunts, warps, or narrows consciousness, limiting man's possibilities of more sensitive adaptation, more resourceful self-determination, fuller growth.

This reading is not, to be sure, biological gospel. In the struggle that now splits civilization, biology, like God, is no doubt on both sides. No more, to repeat, is this a gospel of optimism. If the universe is a creative system, then man cannot expect absolute security; if man is a self-determining organism, then self-destruction is as conceivable as self-fulfillment. The very insistence upon the *potentialities* of man imposes an obligation to allow for constant readjustment, to conceive valuation as not a prescription but a process, to admit experiment, adventure, conflict, and uncertainty as necessary principles. Yet biology does provide substantial reasons for believing that the democratic, humanistic ideals are not mere masks for economic interest, or products of semantic innocence, or symptoms of an effete civilization.

5. SOCIAL BIOLOGY

AN ant colony, W. M. Wheeler maintains, is not a mere conceptual fiction but a true organism. Like a person, it behaves as a unitary whole, resists disintegration, maintains its identity in space and time, even has idiosyncrasies that distinguish it from every other colony. He suggests that as "superorganisms" they are true emergents, and that creative evolutionists might well pause at this social level before skipping on to deity.

> Our biological theories must remain inadequate [he concludes] so long as we confine ourselves to the study of the cells and persons and leave the psychologists, sociologists, and metaphysicians to deal with the more complex organisms. Indeed, our failure to coöperate with these investigators in the study of animal and plant societies has blinded us to many aspects of the cellular and personal activities with which we are constantly dealing.

In such implications of the organismic concepts, biology approaches sociology. Meanwhile it has supplied the basis for at least one revolutionary social theory—the theory detonated by Trigant Burrow in *The Biology of Human Conflict*. Few thinkers appear to have noticed the explosion; but if Dr. Burrow is right, he has shattered the foundations of almost all the world's thought.

He insists that our most revolutionary thinkers have all been superficial in their criticism of behavior. None has had an objective, scientifically established standard of "normal" behavior. All have nevertheless treated society as if they were outside it, free from the symptoms they analyze. Instead, Burrow maintains, the observer must consciously include himself in the group-processes under observation; only so, paradoxically, can he attain true objectivity, as his material is not before him but *in* him.

In other words, we need a much wider frame of reference, including the mental basis of the community itself—the total situation of the racial organism-as-a-whole in relation to its whole environment; and this frame can be supplied only by biology. Burrow and a group of associates therefore carried out an experiment in what they call "phyloanalysis": an effort to get at the fundamentals of biological behavior as well as social conduct, and so to define the *inherently* normal. For some years they lived together, each constantly observing and analyzing himself and all the others, putting all his feelings and findings on the table with complete frankness.

What they discovered, painfully, was a wide gap between intrinsic feelings and the social expressions of them, finally a profound disharmony between the primary, total organismic processes and the secondary, partial system of responses to word-signs. As a result of the extraordinary development of language, man has created an artificial ego, lost touch with his basic needs, forgotten that in the beginning was *not* the Word. He has become so thoroughly conditioned to verbal stimuli, the mere signs of life, that he has become less and less at home with life itself; his constant concern is to adjust the "symbols that *represent* conduct" rather than "the biological processes that *are* conduct." But one symptom of this imbalance is the flourishing profession of education in the "art of love": men today, Burrow remarks, "must be instructed in those functions which in any other animal, however lowly, are performed naturally and without benefit of textbook." Hence the really significant overproduction in the modern world

> consists in the markedly preponderant output of partial meanings and values which rest upon purely artificial symbols and which are produced through the organism's overworked cortical areas in substitution for the total organism's functional activity in its production and consumption of the materials and uses of actuality.

Burrow's own symbols make pretty heavy going. But what he means in general is that modern man literally suffers from a big head; and he also suggests some helpful analogies. Language is properly like a system of semaphores designed to guide trains. Although semaphores are manifestly useful, they do not dynamically influence the train itself. But our symbols or semaphores are now independently organized, in a virtually autonomous system, and treated as if they actually *ran* the train. All the emotional responses called out by these symbols have no connection with the primary motivation of the organism. To shift the analogy, most behavior is biologically as artificial as that of dogs who salivate at the sound of a tuning fork. Conceivably, one could condition dogs to respond arbitrarily to signs made by other dogs—yelps, yawns, tail-waggings. One could then turn loose a lot of dogs so conditioned. Some would accidentally reproduce a stimulus, others would automatically react and thus produce stimuli for reactions in still others, and so one would have a whole dog society whose behavior was elaborately inappropriate, biologically absurd. One would have, in other words, a behaving society precisely like that of man, the lord of the animal world.

An obvious example is the response of men to strictly meaningless oaths—the response of a Southern gentleman, let us say, to the statement that his mother was a dog. But Burrow asserts that *all* our notions of right and wrong, good and bad, are as artificial and irrelevant. The outwardly harmonious behavior of the "good" or "wise" man, the most sincere, unselfish devotion to "justice" and "truth," are biologically as unsound as the obvious disorders of crime and insanity—and in greater need of study simply because they are universally accepted at their face value. All men cultivate a factitious individuality, all seek a symbolic approval from other men, all offer sacrifices to a mere mental image of harmony and health. The more "humane" they are, the more profoundly inhuman they are likely to get. For all these efforts only disrupt the basic unity of the organism as it is reflected in the "preconscious" state: the wordless, selfless state that men still experience occasionally, for example when listening to music. On a larger scale, these efforts disrupt our whole society. The individual sets up certain arbitrary proprietary "rights"; he then comes into conflict with the symbolic rights of others (all men considering themselves right, none more positively than the criminal, neurotic, or social outcast); and this private dissension grows into class conflict, national disorder, ultimately international hatred and war. Altogether, our ability to distinguish between right and wrong, the standard legal test of sanity, is in fact the precise index of our insanity. And so Burrow hopes that his argument gets under the skin, for there is where it belongs.

Here is perspective by incongruity with a vengeance. By this argument Freud is himself a neurotic, Marx a slave of bourgeois mentality, Korzybski a victim of words. All meet concepts with more concepts; all treat only the external signs or symptoms rather than the behavior processes themselves; all share in the profound, universal ignorance of these processes. In other words, all are intellectuals. "Phyloanalysis" offers a scientific basis for the whole anti-intellectual movement, and more especially for the creed of D. H. Lawrence (who wrote an enthusiastic review of *The Social Basis of Consciousness,* an earlier work by Burrow). The organismic processes, the preconscious states, the wisdom of the body that Burrow talks about are precisely the "basic physical realities" that Lawrence spent an impassioned lifetime in exalting and attempting to render in art.

Now an incidental embarrassment of this alarming theory is that it completely disarms criticism. With all respect for organismic processes, one cannot employ them in argument; the critic perforce depends on the symbolic processes, mere words and ideas, and so proves his insanity. For just this reason, however, one wonders whether their kind of scientific Buchmanism did enable Burrow and his group to achieve a "true" objectivity, and whether they did really listen to the voice of the blood, and how they can be sure what it is they heard. Their primary discovery is hardly startling; few will deny that conduct is a continuous effort to maintain appearances and win approval. But when Burrow concludes that this conduct is not a "natural expression of the organism's intrinsic behavior," he has himself been carried a long way by symbolic processes. He has landed, indeed, among pure fictions. For the organism can never be separated from an environment, its behavior is never intrinsic but always a response to environment, and there is no such absolute organismic right and wrong.

Burrow has failed, in short, to comprehend the "total racial situation" he set out to study. This is a social situation, and norms of behavior must be fixed accordingly. Though the efforts to maintain appearances and win approval must involve some disharmony, they are nevertheless as "natural" as they are necessary in this total situation. Burrow has overlooked the inevitable consequences of the evolution of culture. He overlooks as well the necessity of mind; for though he by no means proposes to eliminate it, he is as vague as Lawrence about its proper uses. When he extols the dignity, quiet self-possession, and wholeness of the preconscious state, as it may be experienced in music, he forgets the rich background of conscious experience implied in it and the active intelligence necessary to realize it fully. Children and savages should more often or more nearly experience this preconscious bliss, but if so they do not actually *possess* it; and they are constantly at the mercy of preconscious horrors. There can be no real freedom, peace, and self-possession without the self-knowledge of conscious intelligence, and no rich self to know outside a culture. It is indeed conceivable, as Morley Roberts has said, that the 9,200,000,000 neurons of our higher brain centers may have originated as a malignant tumor, and like overgrowths in other species may cause us to become extinct. If so, there is nothing we can do about it; we cannot cut out the neurons. Meanwhile we might as well make the best of them.

Nevertheless Burrow has something helpful to say about how to make the best. He is attacking real abuses of intellect and real dangers of symbols; we unquestionably do suffer from an imbalance. Those who set a higher value than he upon the exercise of intelligence and the use of the Word have the greater obligation to heed him. Most of them, moreover, do need a broader biological basis for their thought; his specific criticisms of them are usually valid. In general, he does a service by forcing attention on *all* the implications of the organism-as-a-whole—a concept that intellectuals have welcomed chiefly because it solves some intellectual problems.

VI

PSYCHOLOGY

The wider the range of alternatives from which experience may choose, the richer the truth which experience will yield. For experience can only answer questions, and the more varied and intelligent the questions the more illuminating will be the answers. RALPH BARTON PERRY.

1. FACT AND FICTION

IT is not merely flippant to remark that the obvious trouble with psychology is all the psychologists. Strictly speaking, there is no psychology as yet, but only a loosely assorted collection of men who call themselves psychologists. Although these men have discovered many interesting facts and worked out a number of fruitful theories, they have not produced a systematic body of valid knowledge. Indeed, they have not even defined their science or agreed upon its subject matter. Psychology is supposedly the study of mind; but as Carroll C. Pratt points out, no one knows just what is meant by "mind," no one has established a clearly valid criterion for distinguishing the mental from the physical. Conflict arose at the very outset of the new science. Wundt called sensory phenomena mental, and so began to investigate them; Brentano called them physical and ruled them out in favor of the active power of the mind. Progress in the new science has not so much resolved this conflict as multiplied its terms. Altogether, William James's description of psychology is still pertinent:

> A study of raw facts; a little gossip and wrangle about opinions; a little classification and generalization on the mere descriptive level; a strong prejudice that we have states of mind, and that our brain conditions them. . . . This is no science, it is only the hope of a science.

But it is an intoxicating hope, and the extravagance of many psychologists is as understandable as their popularity. To the common man, all opportunity comes at the "psychological moment," no truth is so intimate and impressive as a "profound psychological truth"; it is hard for professionals in such truth not to exploit such moments. Specifically, however, the chief trouble remains the confusion over first principles. As Pratt demonstrates at length in *The Logic of Modern Psychology*, most psychologists seem unaware of what scientific logic is. Their main concepts—perception, trait, instinct, learning, insight, personality, etc.—are usually ambiguous; if they do give these terms a precise meaning, other meanings creep back from the public world. (Hence someone has defined psychology as the subject in which you talk about things which everybody knows about in terms which nobody understands.) When they invent new names, such as libido, racial unconscious, or drive, they are likely to conceive them as entities, psychological facts and not logical fictions. They then think of their statements as literal representations of reality, not abstractions from it, and so are the more inclined to offer a useful but partial description as a complete, essential explanation. Similarly they conceive a scientific law as a commandment enforced by something behind natural events, not a description of these events. And at that their own laws are mostly rough empirical generalizations, rules of thumb.

This kind of innocence especially distinguishes the psychologists who pride themselves upon being strictly scientific. Its most naïve expression is the indiscriminate effort to borrow the magic of mathematics, the rage to measure everything in sight. Intelligence, personality, temperament, character traits—countless things have been tested, graphed, correlated before they have been defined; the figures have some practical value but little scientific meaning.* Sometimes this elaborate technical apparatus is used to discover—or disguise—what everybody knows. A historian solemnly explains that in the investigation of learning "there was a strong suspicion in the minds of psychologists that there might be some significant connection between the learning process and the thinking process"; one result of this suspicion was the "highly significant" discovery that bright children learn more quickly. More often, however, the discoveries have led to a forgetfulness of what everybody knows. Whatever "mind" or "consciousness" may be, it is without question an extremely com-

* An early example of the faith in statistics was Galton's investigation of the efficacy of prayer. On the assumption that sovereigns and children of clergymen are prayed for more often than other people, he studied the records to see if they lived longer. The figures he got indicated that actually they did not live so long. Fortunately, however, the difference was so slight that prayer could not be considered clearly harmful.

plex form of behavior; yet many men still seem to think, as James protested, that no conclusion about human behavior is quite so scientific as one derived from the twitching of a frog's legs—particularly if the frog is decapitated. In general, the natural desire for exactness and objectivity has led psychologists to borrow the methods and concepts of classical physics. They have accordingly slighted the most distinctive characteristics of their own subject matter. Or, like the behaviorists, they have simply taken the psyche out of psychology, denying the existence of all they cannot handle by their methods.

Even the present trend toward dynamic, organic concepts has not cleared up the basic confusion. The new names are still mistaken for things. "Drive," "purpose," "striving," "motivational system"—anything "dynamic" is likely to be regarded as an actual agent, a mysterious force that is inside or behind or more than the events it is invented to describe, much like the ancient "soul" that directs the body and can have a still more mysterious "I" for a captain. Moreover, the reaction against the inadequate descriptions of mechanistic psychology has carried back to what might be called the literary conception: the idea that the descriptions of psychology must always agree with the immediate feel of consciousness, express the vital "human" quality of experience. Other scientists, K. S. Lashley protests, are never obliged to arouse the experience of the thing described:

> No one asks that the physicist's account of gravity shall make his hearer feel heavier, or that the biologist shall throw him again *in utero* by his statement of the recapitulation theory. Yet many psychologists demand that the explanation of mind shall be, somehow or other, identical with mind.

The common complaint is that the psychologist's account is incomplete and inhuman, misses the "living truth" of experience. The answer is that any properly scientific statement, as a severe abstraction, must be incomplete and inhuman. To render the "living truth" is the business of the poet, not the psychologists.

A good illustration of these difficulties is Gordon W. Allport's *Personality: A Psychological Interpretation.* Allport soundly insists that analysis must take into account the fact of complex patterns and the fact of unique individuals. He protests against the view that personality is merely a bundle of attributes, individuality merely a particular combination of universal traits or drives. In any concrete sense, he declares, there is no such thing as intelligence, space perception, honesty, sexuality; there are only unique individuals capable of such things. One could say as much, however, of the concepts of physics. They all involve abstractions or fictions, such as "mass" and "energy"; yet they enable an accurate prediction of how all the unique physical objects will behave. Again, Allport warns against the concept of "types" of individuals and of "common traits" shared by them—generalities that can be roughly useful but are never "real." His basic element of personality is the "true trait," the unique pattern of the individual. But although this may be a more adequate concept, it is still an approximation, a convention; in so far as the "true trait" can be defined and strictly known at all, it too is a generality. In short, Allport's desire to do complete justice to the individual has given him an excessive fear of all abstractions—except perhaps that of "individuality." At times he forgets that a scientific statement is necessarily general, fractional, arbitrary, and can never do complete justice to the individual. Psychologists can and should recognize uniqueness; only artists, once more, can express the unique object.

All this is to say the worst about psychology. I say it at the outset, deliberately, emphatically, for the confusions are serious. Yet I say it to get it out of the way, for one can also make too much of it and then overlook the obvious. Psychology remains a most important branch of inquiry and has already made significant contributions. Its facts cannot be ignored because they have not yet been adequately formulated, nor its theories because they are not yet in perfect scientific trim. Furthermore, the mysteries of "mind" do not justify the conclusion that psychology can never be a respectable science. Whatever the relationship may be between physiological events and conscious experience, all observation indicates that it is constant and orderly, and therefore susceptible of strict scientific formulation. If mental processes are difficult of access and control, they are not taboo because subjective, or even when accessible only to introspection. All phenomena are private to begin with. They become objective, Pratt points out, when they are reported; and whatever can be adequately reported is a proper subject matter for science. In the final analysis, the observation of a salivating dog is as subjective as the sensation of a bellyache.

The layman, at any rate, need not worry too much about the

canons of scientific truth. The severe logician is displeased by the many psychologies, and sees in them chiefly the need of unifying concepts, a common base of operations. The literary critic can be content with the obvious pertinence of all these psychologies, and can even welcome their diversity. In adapting the various provisional findings to his own uses, he can get along well enough with practical wisdom. He need not wait upon the exact formulation and the comprehensive concept; he need not let it matter just what "mind" is or mind just what "matter" is. Mind may be conceived in purely physiological terms as the brain, and its processes studied accordingly. It may also be conceived, in Dewey's terms, as "the body of organized meanings by means of which events of the present have significance for us." Psychology may set as its final task the correlation of these processes and these meanings. Meanwhile psychologists have profitably investigated both. They have thrown light on the immediate perceptions that are fundamental in all thought and behavior, and on the needs and interests that direct thought and behavior. The critic can be the more grateful for any light from any source because in his business he cannot expect the kind of truth demanded by severe logicians.

In other words, he will be an eclectic. But he need not feel like a dilettante in this role, for he will be in the company of most psychologists. Many do not consider it necessary to identify themselves with a definite school; many more who do label themselves admit the legitimacy of other schools and feel free to borrow from them. Thus the great majority of psychologists now accept both behavior and consciousness as proper subject matter—at least when consciousness is not conceived as an immanent something in addition to experience. Similarly they no longer feel obliged to choose between structuralism and functionalism, recognizing that structure implies function and that function is impossible without structure. And even when they are very exclusive, making gospel of theory and damning the other fellow's theory, their underlying agreements are more significant than their differences. Most psychologists now conceive mind as an agent that pursues its own purposes, not a mere container of ideas or a mechanism that merely reacts to stimuli. (A century ago, Victor Branford has pointed out, a child's mind was "filled" or "polished"; now it is nurtured or trained, like a plant.) They conceive experience as a dynamic organization, not an aggregate of elements linked by some purely mechanical process of association. These terms are still vague and the old habits of thought still stick; in their piecemeal analyses psychologists are apt to fall back unawares upon atomistic, mechanistic assumptions; but most of the schools are united in their stress upon the whole personality behind the momentary consciousness, the whole organism embracing the parts isolated by analysis.

As an outsider, the critic can more easily keep clear of the unnecessary controversies that separate the schools and obscure their contributions. He can follow the sensible policy, again, of welcoming the positive discoveries and discounting the negative conclusions. And again this is not a carefree or slapdash procedure. Practical wisdom calls for discrimination as well as hospitality, and in this field more reservations than usual. In particular, the critic must be wary of the literary criterion of psychological truth.

> There are two portraits of human nature [Pratt summarizes]: one given by intuition and direct description, the other made by systematic and experimental analysis of the conditions correlated with the events initially described. The two portraits do not agree, for they are not the same thing. The first portrait is merely a point of departure. The second portrait is the major concern of psychology.

The first will remain the major concern of literature, and will continue to have its own kind of validity and usefulness; but it is no more the necessary test of the second than our immediate perception of a table is a test of the physicist's picture of a whirl of electrons. Literary men commonly praise and condemn psychology alike for the wrong reason.

2. PHYSIOLOGY AND BEHAVIORISM

"THE brain," wrote Arnold Bennett, "is a servant exterior to the central force of the ego." Although the question arises of where, then, this ego is located, and how it gives or enforces its orders, this naïve dualism is engrained in language and common sense. We "make up our minds," we "use our heads," we "cudgel our brains"—and it is indeed convenient to talk as if there were an ego in charge of everything. Nevertheless we are not forced to take such talk seriously; this ego is mystical only if we choose to make it so. It is finally inconvenient to go out of our heads.

For the brain is a fact, no mere fiction. Its close relationship with mind or consciousness is unquestioned. Whether or not psychology should, as many believe, move always closer to physiology, this is one obvious place to begin—especially since inquiry here can be scientific by the strictest canons.

Specifically, we begin with the whole nervous system. This may be pictured, roughly, as a series of integrated levels, from the visceral centers through the thalamus or lower brain centers to the cortex or "gray matter." (I am here drawing chiefly upon the works of C. Judson Herrick—*Brains of Rats and Men* and *The Thinking Machine*.) Thus the thalamus, which is the center of simple emotion and supplies the immediate qualities of perception or sensation, regulates and reinforces the elementary visceral happenings, discharging downward toward the motor centers and upward toward the cortex. The role of viscera in emotional experience has long been recognized in folk language, which has made the heart or the bowels the seat of emotion and today stresses the importance of guts. The role of the cortex in such experience has also been recognized, but is seldom fully appreciated. The thalamus of chickens is nearly as good as that of man; man has a far greater capacity for feeling as well as thinking because of his cortical operations. By the same token, however, we have to distinguish between "genuine" emotion, with visceral reverberations, and the imitation or illusion of emotion produced by the cortex. Students have sincerely reported an "emotion" at the mention of the word "mother," although a galvanometer indicated no bodily change whatever. They have also reported no emotion at the mention of "prostitute," although the galvanometer gave a definite kick; here the emotion had been suppressed at the higher conscious level.* Literally, we may not know what we are feeling. Hence the "subconscious mind" can be given a respectable neurological status; unconscious mental activities are real body functions.

In this picture, the first thing to be emphasized is the integration of the whole nervous system—the organic interdependence of thought and feeling, brain and viscera. The most abstract idea grows out of sensory excitement, the loftiest ideal out of raw

* Such illusory or imitative emotion obviously enters esthetic experience and complicates critical judgment: much appreciation or condemnation is primarily cortical. This is another way of showing how and why doctrine can blind a critic to his actual esthetic experience, or can inhibit such experience.

emotion; all thinking and feeling are biological processes, which, like breathing and digesting, involve the whole organism. Most important for man, however, is the distinctive role of the cerebral cortex. All protoplasm can learn; the amoeba is an agent of selection, a creature of purpose; the ape has the ability to solve problems. But when these activities are consciously directed and controlled, a strictly new kind of behavior emerges. It is this difference in kind, as well as in degree of complexity, that gives psychology its own subject matter and makes it more than an incidental branch of biology.

The general function of the higher brain centers is to provide an intelligent guidance of behavior. Reflex adjustments take care of many events and are adequate for ordinary occasions; the cortex steps in when these simpler systems cannot determine the appropriate response, or interfere with one another so as to inhibit response. In Herrick's terms, it regulates behavior in process by "first checking inappropriate reflexes and then amplifying, redirecting, recombining, or otherwise improving upon the immediate responses which have already been initiated in the subcortical reflex centers." In a real sense a man stops to think, makes up his mind. Herein lies the importance of "inhibition." An infant, as any parent knows too well, has no inhibitions; if it responds at all to a stimulus, it responds at once and with all it has. The adult holds back to discriminate, decide what the situation calls for, and so can give a measured response. Hence we need to guard against the misleading connotations of the term. "Inhibition" suggests merely negative behavior—a suppression, a stoppage; whereas it is a mode of action, the means to positive decision. Since Freud it also suggests an unhealthy suppression of vital impulse, a stunting of the lovely young idea—as indeed it can be. But it remains the necessary means to maturity, poise, and self-control.

It is only the first step, however. The most distinctive power of the cortex is mnemonic. Somewhere and somehow in the course of evolution, the organization of brain tissue became complex and plastic enough to enable man to learn rapidly and retain his learnings in abstract or symbolic form. The cortex now holds up the train of reflex behavior in order to assemble and discharge the results of past experience. It contains, says Herrick, the "living residue of past experience," "structural alterations which form the organic basis of our memories of former events and our reactions to them." His physiological description of the brain accordingly comes pretty close to Dewey's definition of mind as a "body of organized meanings." Also important, moreover, is the measure of autonomy of the higher brain centers. A train of thought is started by some happening relayed through the lower centers, and may at any moment be derailed by subsequent happenings, but it has its own tracks and semaphores. Although its course may be a sequence of unconscious or involuntary associations, it may also be consciously redirected by the demands of *logic*—which is not a matter of sensory or visceral excitement. Science itself is possible only because thought and behavior are not so automatic as many scientists have assumed. Altogether, the cortex has its own purposes and necessities, not independent of the whole organism, not "uncaused," but not wholly dependent upon or caused by the more rudimentary physiological processes. It functions much like the elected representative of a people, who is governed by their wishes but has his own parliamentary rules, and who also governs the people and influences their wishes.

In *A New Physiological Psychology,* W. Burridge objects even to the suggestion of anatomical dualism in this concept of "higher" and "lower" centers. On the basis of his experiments on hearts, he has worked out the theory that the whole nervous system is a rhythmical structure which, like the heart, is always in activity, rather than a series of parts which, like the supposedly quiescent muscles, must be stimulated into activity. It is a colloidal system with two sources of potential energy; what have been called its functions are "manifestations of kinesiphore activities," the relations between these two sources. The details of the theory are highly technical. The general idea, however, is that physiologists have hitherto observed the total activity of the system but not the component actions—the marriage but not the partners—and so have misinterpreted "normal" expenditures of nervous energy, in which actually the extravagance of one partner might have been balanced by the strict economy of the other. The ultimate source of variations in behavior is a quickening or slowing of rhythms, an augmenting or diminishing of amplitudes; "wisdom" is apt to be sad because the rhythms of age are stable but also slower than those of youth. The immediate source of variation is a change in relative potential, always on the principle that the more energy the one partner supplies, the less comes from the other. Yet this principle of compensation or balance does not imply a conflict, such as Freud assumed between the pleasure-principle and the reality-principle. The relation is instead like that of whiskey and soda in a glass of given size (the total energy-capacity of a given organism). The more we pour of one, the less we can have of the other; but the result is never a victory for either, it is always a whiskey-and-soda. Altogether, the principle of rhythmical structure implies a still closer integration, a still wider harmony. "The whole beats warmer than its previous parts." *

As I understand it, this theory might be supported by the experiments of Northrop and Burr on the electro-dynamic field of the organism. Burridge often draws upon electricity for his symbols; he explains the interdependency of neural and psychic processes—or in other words of matter and mind—by a kind of "transformer" in the brain. Meanwhile, however, this highly technical language is also figurative; and it is well to mark clearly the limits of our present knowledge. The exact relationship between neural processes and psychological processes—the data of actual *experience*—is unknown. Ultimately even the physiological happenings are enigmatical. "We do not know," Herrick notes, "exactly how a sense organ is excited, how a nerve fibre conducts, how a muscle contracts, how a gland secretes, or how the brain thinks." Although we have satisfactory evidence that these organs do perform these functions, we cannot explain the mechanisms we infer. The technical terms help to locate and define the problem but do not solve it—"protoplasmic irritability" is not much less mysterious than *élan vital.* Altogether, as I. A. Richards himself admits, the scientific account of the nervous system is so far "only a degree less fictitious than one in terms of spiritual happenings." John Crowe Ransom goes further, maintaining that at present "the mental datum is the fact and the

* Burridge suggests a physiological basis for our pleasure in scientific theories. Logically they may be only shorthand descriptions, but psychologically they are *harmonizations* of the facts. Facts do not please us when disconnected and seem less important once they have been integrated; the fact-process and the theory-process are also "kinesiphore activities." Hence our greater pleasure in comprehensive theories, the supertheories of theory, the approaches to the "ultimate reality" of philosophers. They are likely to be questionable and unlikely to enable better control of the environment, but they make still wider connections, create still larger and more harmonious wholes. This pleasure, I should add, points to the psychological affinity between the scientific theory and the work of art.

neural datum is the inference." And for literary criticism the mental datum is no doubt still the more useful metaphor.

Nevertheless there remains a general correspondence between the fact and the inference, and what we do know of the nervous system is suggestive. Above all, there is no necessary opposition, nothing incompatible between our most cherished values and our knowledge. The "inner check" and the "stream of impulses," wrote Paul Elmer More, are "essentially irreconcilable" forces. "What, if anything, lies behind the inner check, what it is, why and how it acts or neglects to act, we cannot express in rational terms." Physiologists express just this, in altogether rational terms, and they find no such irreconcilable opposition. They conceive intelligence, abstraction, idealization, and all these "higher" activities as part of the biological mechanism of control; if they do not completely explain this mechanism, at least they do not make a needless mystery of it or split us in two. "Voluntary self-control is effective," Herrick concludes, "because it is no penumbra of an ethereal spiritual presence floating around and into our personalities from the outside void." It is a real power, located in the cortex, and like other organs the cortex is capable of doing real work.

At any rate, this account of the nervous system is in line with the main trend in psychology today, toward functional, organic concepts. It supports the view that analysis must also take into account the more complex levels of behavior, and finally the organism as a whole. And it is the more pertinent as an introduction to the school of psychology that prides itself upon being the most physiological and scientific—behaviorism.

Given all the foggy idealism that has been inspired by the idea of "consciousness," not to mention all the trouble it has caused in psychology, one can understand the behaviorist's wish to get rid of the idea for good by simply denying that there is any such thing. Nevertheless this is a conscious wish, and the behaviorist must know that he not only behaves but is aware of behaving. To be sure, one cannot absolutely prove that other people have minds of their own and know what they are doing. But all argument is futile except on the assumption that they do, and in their own arguments the behaviorists perforce make this assumption. They join the search for "truth," they are rightly proud of their important contributions to "knowledge," they do their best to change other people's "ideas"—and they prove that behavior is automatic by the curious procedure of conditioning behavior to suit themselves. Their whole performance is the more curious because it ends in mental hygiene: a self-conscious effort to disprove consciousness becomes a conscientious effort to teach people how to behave. For scientists, the behaviorists are very free with the notion of "ought," and seem unconscious only of the norms it implies.

The root of their confusion is again an old-fashioned scientific innocence. The behaviorists have failed to learn that the serious business of psychology, as of every science, begins at the level of concepts, not of raw facts, and that the concepts must be derived from the subject matter, not imposed upon it. For they are among the last survivors of the good old days, when all the world was real stuff, stuff without nonsense, and good sense lay in recognizing the unreality of all sense. They believe that they are ultrascientific because, like classical physicists, they have abstracted from their subject matter all its distinctively human qualities—the awareness of behavior that constitutes "experience." They believe that they are men of sterner stuff because they grimly hang on to an unpleasant but elementary fallacy, that since consciousness may be reduced to bodily movements it is identical with these movements. The faith that the complex may be wholly explained in terms of the simple, or that the simple is more real because it conditions the complex, is the tender-mindedness of science. It finally lands the behaviorists in an impossible idealism.

Nevertheless it is easy enough to disregard their extravagances, especially since Pavlov, the great pioneer of behaviorism, never claimed or denied so much as have John B. Watson and other American followers. Nothing in their positive findings requires the denial of consciousness, all the findings are as pertinent when consciousness is admitted. Forget their mechanistic philosophy and their mechanisms still work, to more purpose. Pavlov's experiments have thrown light on the delayed responses or inhibitions studied by physiologists. They have been taken over by the Gestalt psychologists, who for years have been battling with the behaviorists; in their own experiments with animals, they also condition behavior but show that the reflex is touched off by the perception of a whole, elements-in-relation instead of an element-in-itself. (Chickens who have learned to go to the smaller of two pans for their food will ignore this pan when it is made the larger of a new pair, and head at once for the still smaller one; what they have perceived and learned is a relationship, not the pan itself.) Likewise the conditioned reflex helps to explain the Freudian complex—the mechanism if not the motive behind neurotic behavior.

Hence the best reason for ignoring the philosophy of the behaviorists, or at least dismissing their pretense that it is not philosophy but pure science, is that one may then do more justice to their notable contributions. They have introduced a healthy emphasis on what people actually do—the behavior itself instead of the endless talk about it. They have provided a most useful principle for the understanding—and possible education—of all social behavior. If one may still question Watson's effort to reduce language and thought to subtly conditioned muscular activity, there is no question that the conditioned reflex has much to do with the processes of abstraction and above all of identification, the responses of men to names and symbols. It is behind the "stock response," the fixed principle in literary criticism as in all other thought. In general, the notion of conditioned behavior has become an indispensable part of our intellectual equipment. If it is another way of saying "habit," it is a more precise way, and more fruitful in that it makes possible a fuller understanding, suggests further ways of doing something about it.

3. PSYCHOANALYSIS

Sigmund Freud could write, in all modesty, that he belonged with those who have "disturbed the world's sleep." His work has made a permanent difference in man's thinking about himself; the world will never again sleep quite so soundly, or consciousness regain its innocence and purity. The idea of an unconscious mind is very old, of course. In the literature and philosophy of the nineteenth century it is even conspicuous; Schopenhauer in particular came close to the id in his concept of the Will, whose "pole or focus" he located in the genital organs. Yet before Freud the idea had never been clearly grasped or fully realized. He made it explicit, systematically worked out its implications, developed from it a remarkably fruitful set of theories. His basic contribution is as original as it is incontestable, and beyond the power of criticism to destroy.

For this reason, however, criticism is the more necessary. Freud's main principles are so familiar and so generally accepted today that they no longer require explanation or defense; their usefulness has been thoroughly demonstrated, both in practice and in theory; but their limitations—apart from his exaggeration of the role of sex—are not so clearly or fully realized. We now need to dwell on these limitations, simply because psychoanalysis is so important a tool.

To begin with, Freud's hypotheses have as yet little strictly *scientific* use. Their very richness indicates not only that they are questionable, but that they have not been so formulated as to permit effective questioning by scientific methods. Thus the various schools within psychoanalysis continue to battle over conflicting interpretations; but they have performed no critical experiments to decide among them, nor even tackled the problem of how to devise such experiments. They admit so many variables in their interpretations that the analyst can discover almost any "latent content" he has a mind to. Freud declared, for example, that the meaning of a dream element may be what it appears to be or just the opposite, it may be taken as a memory or as a symbol, or its significance may lie in its wording. The impulse behind it may be an unsolved problem, or a task accidentally left incomplete during the day, or a suppressed impulse, or an unconscious impulse excited during the day, or just some indifferent impressions that are therefore unsettled. As for symbols, neckties, umbrellas, women's hats, ladders, nail files, revolvers, ovens, bridges, hammers, cupboards, snails, mice, sticks, and countless other objects serve for the genitals or the sexual act. Having all these cards to play with, the analyst can take any trick—but the game is scarcely scientific. However ingenious or illuminating his interpretation, it has little in common with analysis as the chemist knows it.

Even the undeniable therapeutic successes do not necessarily prove the truth of psychoanalytic theory. Erroneous theories may account satisfactorily enough for the facts at hand (as they have throughout the history of science), "cures" have also been worked by religious and magical means, and psychoanalysts do not really know why their own cures often fail to work. Nor have they sufficiently considered such logical problems, the criteria of scientific truth. It is significant that they have not tried, until recently, to get in touch with the rest of psychology, or in line with the main march of science. They have made little use of the knowledge accumulated about learning, forgetting, desiring, striving, etc. Their neglect of physiology—the neural mechanisms to carry out the operations demanded by their increasingly elaborate metaphors—is a striking example of an isolation that no science, least of all a youthful one, can afford.

In general, the account of the mind rendered by psychoanalysis is more like a literary description than a scientific formulation. At worst, its animistic terms—libido, ego, Censor, Life and Death instincts—encourage the yogi tendency in Freudian therapy. At best, they are still vague and apt merely to conceal problems. "Sublimation" may satisfy a logical requirement, but it may or may not denote a real psychic process, and in any event it does not explain why the sexual impulse should lead to the art work instead of the sex act. If one takes the whole story literally, troublesome questions arise about just who or what is being satisfied or fooled by all the fantastic subterfuges, and how the not too bright Censor got and holds his job, and what kind of satisfaction an individual gets from symbolically satisfying a desire he doesn't know he has. If one interprets the story freely, the strictly scientific questions do not arise at all. When "libido" is conceived as sexuality (as in practice it usually has been), it explains considerable but has to be strained beyond belief to explain all it is supposed to. When it is broadened into Eros or the Life Principle, it becomes mere tautology and explains nothing. And all this leads to a serious practical criticism—the failure of psychoanalysts to define their standard of normality or mental health. Their whole practice is designed to "adjust" people to "reality"; Henry Miller asks the pertinent question: *what* reality? *whose* reality?

Specifically, the limitations of Freud's thought derive from the attitudes and the concepts he took over from the rationale of the nineteenth century. Although he was at bottom modest and cautious, at one time or another making all the necessary disclaimers, he had the habit of offering his statements as "incontestable," "unmistakable," "self-evident," the "only possible explanation"; and this dogmatic habit was called out not only by the bitter opposition he met, but by his belief that scientific truth *was* self-evident and a scientific theory that apparently worked *must* be the only possible explanation. He confused the assumption that there is nothing arbitrary, accidental, lawless in mental life—an assumption indeed necessary for a science of psychology—with the quite different idea that there is nothing arbitrary in scientific propositions *about* mental life. In this absolutist spirit he asserted that the unconscious is the "true psychic reality"; and he would have fiercely denied that there is anything metaphysical in this proposition. He was unaware, generally, that his whole reality was based on questionable premises. And because he was also an actual pioneer in thought, these premises are confused and inconsistent at that.

Freud brought some new principles into psychology, concepts of purpose, dynamic relation, organic continuity; he had discovered that neuroses were purposive and active, and that an understanding of them required a knowledge of the patient's entire life history. At the same time, he never quite got away from the old-fashioned dualism of matter and mind; he declared that "*psychic reality* is a special form of existence which must not be confounded with *material reality*," and he claimed independence for psychology from all other sciences. Nevertheless he went to biology for his ultimate psychic fact—the instinct. This old-fashioned theory of instincts led to further dualisms, for to explain some purposeful behavior he felt obliged to invent a Death instinct; Anna Freud speaks of man's instinctive antagonism to his instincts—leaving open the question of why nature should select and breed such a two-headed monster. The notion of fixed elements of universal human nature in turn leads back to the mechanistic scheme of physics, which explains Freud's habit of thinking in terms of psychic atoms. As Dr. Alexander R. Martin points out, he stressed the noun rather than the verb in mental life. Characteristically he talked of the unconscious as a "storehouse," a submerged place, and of the mind in general as filled with symbols, ideas, memories—psychic *things* in space instead of movements, tendencies, psychic *events* in time. Altogether, he failed to take a consistently dynamic, organic view.

The most unfortunate of these legacies was the theory of instincts. It not only landed Freud in a profound fatalism, which was inconsistent with his faith in knowledge as a means of control, but caused him to neglect the very important influence of culture. He did not see that the unconscious as well as the conscious mind—the impulse as well as the repression—is socially conditioned, and that the "normal" manifestations of sex itself are shaped from the outset by social conventions, such as the modern convention of romantic love. For the same reason he did not see that motives are also individualized. Biologically, sexuality may be more or less simple and uniform, but psychologically it is highly complex and idiosyncratic, tied up with a system of interests, traits, habits in which it is not the "cause" or normally even the dominant force. (That it seems to be the dominant force in most neurotics would prove only that they *are* neurotic.) In short, Freud analyzed only a fraction of the total personality. He misinterpreted the whole because he tried to explain it in terms of this part; he misinterpreted the part because he ignored its organic relation to the whole. And the practice of psychoanalysis has led to further distortions. Its technique is frankly therapeutic and can handle only repressions and frustrations; even in theory it does not yet apply to healthful, vigorous, joyous, creative activities. Yet it has prejudiced men's conception of the whole mind and personality, much as their idea of the stomach would be distorted if they conceived its natural function to be indigestion.

It is the more necessary to stress these limitations because the official disciples of Freud have notoriously bent their efforts to expounding the true gospel. In his introduction to the *Basic Writings* Dr. A. A. Brill, for instance, describes all criticism as "defection," and adds that of course "we have nothing to do with the so-called analysts of the other schools." But the early heretics did not improve the text. Although they offered sound criticisms, they wanted to usurp rather than qualify the absolute authority of the master; they made gospel of another useful but inadequate set of principles. These different principles actually do not seem irreconcilable—to the outsider the warring schoolmen all look like psychoanalysts. Nevertheless they remain schoolmen.

Thus the "individual psychology" of Alfred Adler is an ambitious effort to explain *all* behavior, conscious and normal as well as unconscious and abnormal. He objects to the dualism in Freud; he interprets behavior as an organized striving toward a definite goal, in which consciousness and unconsciousness are not antagonists but different means to a common end. But like Freud he insists upon a single, absolute causal principle. He replaces the sexual drive with the "will to power," and so makes the inferiority complex the source of all neurosis. The outsider, again, wonders why both principles cannot be retained, for both seem pertinent, neither seems sufficient. Adler points to vital purposes that cannot be explained by sex, and deplores Freud's stress on endless conflict because it permits no real peace or harmony; Freud points to behavior that cannot be explained by a will to superiority, and deplores Adler's stress on the impulse of aggression because it leaves no room for love. Jung, on the other hand, leaves room for practically everything, but only by having a precise place for practically nothing. He defines the mainspring of behavior as neither sexuality nor the will to power but an undifferentiated life-energy that can find innumerable modes of expression; he defines the unconscious as a storehouse that contains not merely the individual's suppressed experience but the memories of his family, his tribe, his race—mythology, folklore, history, the arts, the whole drama of life on earth. In this immense vision all symbols become cloudy, splendid for poetry, perhaps, but useless for science. Scientists can make nothing of the "racial unconscious," at least when it is conceived as something more than what the individual learns from the group or shares with it because of common biological purposes; they have never found consciousness without a specific nervous system, they cannot hope to find a physiological location for a group-mind. And literary men, like Jung himself, are prone to make too much of it, conceiving it as a separate entity and then a superentity—the universal mind, the Oversoul.

Far more significant, accordingly, is the movement recently started by such psychoanalysts as Karen Horney, Abram Kardiner, and Alexander Martin. They are bringing Freud's thought into line with the main developments in psychology and the sciences generally, reclaiming it in more inclusive concepts. They interpret behavior as a "psycho-biological totality" whose fundamental force is "integrative"; the individual acts and reacts as a whole, tending to "hold on" or "tighten up" all along the line—somatic, psychic, social. Of this total integrated pattern, sex or the will to power is only a part, not the essence or determinant. Neurosis they trace back to a "basic anxiety," a "feeling of being alone and helpless in a hostile world"; the mind loses its own integrity when it loses integration with the environment. The Œdipus complex or the neurotic wish for power is more a symptom than a cause of this anxiety. But both the source and the sign of neurosis are cultural, not simply instinctive or constitutional; the "cause" is finally the whole situation. This whole departure, Dr. Martin maintains, is along roads opened by Freud in his later work, though not explored by him. It is at any rate a more impressive tribute than is the worship of the Freudian fundamentalists; that his basic principles can be readapted is more noteworthy than that they have first to be overhauled.

Even more significant, however, is the growing disposition of psychoanalysts to question the *primacy* of the unconscious. Freud himself never glorified the id. His aim was always to control it, dominate it by reason; an old-fashioned rationalist, he aspired to a pure intellectual virtue, the "heavenly city" of eighteenth-century philosophers with modern improvements. "Where Id was, there shall Ego be." Yet he clearly made consciousness secondary, regarding it as an imperfect "sense organ" for the perception of the "true psychic reality." In his account, the dark powers of the unconscious always play the leading role in behavior. Hence he left an opening for the most serious abuse of his discoveries: the widespread contempt of intellect, the worship of the instinctive and the irrational. No paradox of modern thought is more tragic than this, that the discovery and analysis of the unconscious, which is a triumph of conscious intelligence, should come to be viewed as a humiliation of intelligence, a reason for idolizing the dark powers.

Similarly, Freud's analysis of unconscious motives has contributed to the widespread distrust of all conscious, avowed motives. The great job of psychology, said Nietzsche prophetically, was to expose the "innocence" of the universal intellectual dishonesty. It was indeed an important job; rationalizing has always been far more prevalent and more dangerous than faulty logic. But the job has been done at once too well and not well enough, destroying the innocence but not the dishonesty. Psychoanalysts themselves, in the first place, can too easily discredit any serious criticism of their theory: opposition betrays a secret "resistance," automatically commits the critic to the position of trying to evade disagreeable truths. Apart from all trickiness, however, the descriptive terms of psychoanalysis are freighted with unnecessarily disagreeable connotations. "Maladjustment" suggests weakness or unfitness, "compensation" a mere shift or fraud, "repression" a timid or unhealthy denial of normal impulse. As a matter of fact, the sensitive, acute, or ardent man—the man who for any reason is superior to the group—is for this reason apt to be maladjusted; compensation is often a very sensible procedure, and may lead to the greatest achievements; a good deal of repression is indispensable to the normal functioning of intelligence. In general this vocabulary can confuse the issues of honesty, obscure the proper means and ends of intelligent behavior. In creating the illusion of going very deep, it has plainly led to much shallowness.

This kind of shallowness has nowhere been more apparent than in the psychoanalytic approach to literature; and here Freud himself is more directly responsible. Although he had a natural respect for the artist, he could not, as a scientific rationalist of the old school, entrust to him any of the really serious business of life. "So, like any other with an unsatisfied longing," he said, "he turns away from reality and transfers all his interest, and all his libido too, on to the creation of his wishes in the life of phantasy, from which the way might readily lead to neurosis." If art is almost always harmless, or even useful as a "narcotic," the reason is simply that it is accepted as pure "illusion": save for "a few people who are, one might say, obsessed by Art, it never dares to make any attack on the realm of reality." (These few people, one might add, include all the great artists of the past.) At the same time, since Freud's criteria of the real and the normal were not clearly defined, his conception of art was also ambiguous. He often admired literary artists because they had anticipated so much of his own science of the unconscious; in other words, these specialists in "illusion" had had profound insights into the "true psychic reality." He admitted that "substitute gratification" is often valuable in itself as well as practically necessary, but he also stigmatized it as an inferior substitute, an evasion; and just when it is one or the other in art he did not say. Yet even the "good" gratification in this kind is a hopelessly inadequate description of great literature. *Œdipus Rex, King Lear, The Brothers Karamazov*—to account for such expressions of deep and terrible experience in terms of wish fulfillment, narcotic, illusion, a turning away from reality, is simply to lose one's grip, to escape the responsibilities they impose.

Hence the obvious pertinence of psychoanalytic theory for literary criticism makes it necessary to mark clearly the limits of its application. As Freud himself admitted, to begin with, it throws no light on the creative urge, the impulse to impart form, or on the artistic gift that enables this impulse to be handsomely gratified. Consequently it throws no light on the artist's peculiar effectiveness—the qualities that distinguish a first-rate from a third-rate work of art, and any work of art from other forms of substitute gratification. Even if we admit Freud's questionable theory that the esthetic sense derives from sex, or that "the beginnings of religion, ethics, society, and art meet in the Œdipus complex," we still have only a partial explanation of how and where art begins, no light on why and where it ends and what makes it end well. In *Expression in America,* Ludwig Lewisohn constantly explains the limitations of American writers by deficiencies in their sexual powers or frustrations in their love lives, and in particular finds mother fixations everywhere; but since he spots these in writers he admires, such as Dreiser, and does not make them the source of Dreiser's limitations, we can never be sure just what he is explaining. And here we run into the positive confusion—of conditions with causes, origins with essences—that has muddied this whole approach to literature. Dr. Brill makes the remarkable statement that the love of poetry is simply an expression of "oral eroticism," "a chewing and sucking of beautiful words." Others are content with less ingenious variations on the main theme, that because art satisfies some wish it is nothing but wish fulfillment.

The gist of all this is that psychoanalysis is most pertinent in studies of the temperament and behavior of the artist as citizen, not as artist. The information it gives is interesting and useful, as is all biographical and historical material; it contributes to a fuller understanding of his work as an artist. Nevertheless it is always marginal, like the scholarly footnote. The artist's private life is as much the effect as the cause of his work, but in either aspect it is not a master key. Countless writers have been maladjusted or neurotic—and all have written differently. In countless ways they may project, symbolize, sublimate, or compensate for their frustrations, or like Proust they may even directly represent them; in any event the kind of mess they make of their lives does not determine the quality and value of their work. Hence the critic needs to be wary as well of the "inner meanings" that psychoanalysis discovers in works of art. It can discover such meanings, since art contains dream elements and is a mode of self-expression. It can also force these meanings. Usually the interpretations are too schematic and stereotyped: the sun must always symbolize the libido, anything erect is erection, all that meets the eye is the genitals in disguise.* But at best these hidden meanings are not to be taken as the "essential" or the "real" meaning of the work, since art is not only dream and self-expression but conscious graph and communication. When Dr. Ernest Jones saw an Œdipus complex in *Hamlet,* he may have got hold of something, perhaps even the secret of Shakespeare's unconscious intention. He did not get hold of the play and the poetry, the wealth of implication, the secret of Shakespeare's *success.*

Finally, however, the most important issues for literary criticism center about the role of the unconscious. It is indisputably a primary role in art. That "inspiration," "vision," the "magical power of imagination" cannot be summoned at will, issue from a self not consciously known, has been recognized ever since the poet began invoking his Muse, and Plato described him as one "not in his right mind." Yet as indisputable is the necessity of the conscious mind for recognizing the inspiration, shaping the vision, wielding the magical power. The Surrealists' effort to render literally the "pure" unconscious is itself a conscious intention, and usually all too self-conscious. This purity, moreover, is a fiction. What comes up out of the depths is some amalgam of past experience in the waking world which contains ingredients of rational thought; the dreams and imaginings of a savage or a child are different from those of a civilized adult. It is impossible actually to separate the two minds, the sleeping and the waking worlds. But the practical purposes that justify the convenient distinction also force us to regard the unconscious mind as subordinate, only a means to some conscious end; for it can be thought about and known only by the conscious mind, its uses can be realized only in the waking world. It is pointless consciously to argue the superiority of the unconscious. And here is the issue.

The romantic poets and critics—Coleridge, Wordsworth,

* In his often fascinating *Problems of Mysticism and Its Symbolism,* Herbert Silberer analyzes a medieval hermetic parable. The unconscious intention of an elaborate narrative comes out so: "The wanderer in his phantasy removes and improves the father, wins the mother, procreates himself with her, enjoys her love even in the womb, and satisfies besides his infantile curiosity while observing the procreative process from the outside." Silberer adds that "possibly one may be surprised" at this whopper, but explains that the imagination has "titanic powers." More surprising is the narrow range and infantile interests of these titanic powers.

Carlyle, Emerson—who spoke of deep springs beneath intellect, which are the springs of poetry, spoke of them as sources of wisdom and power. This was a way of reaffirming the higher truth of poetry, and perhaps of protecting it against the competition of science. In *The Poetic Mind,* Frederick C. Prescott systematically worked out this way, in one of the first notable efforts to relate Freudian theory to a theory of poetry. "On the whole," he maintains, "the unconscious mind is superior in insight and wisdom to the conscious one; and this must be insisted upon because it will demonstrate the superiority of poetry." The unconscious is older, deeper, richer; it can take a "wider view" because it shares in the "mind of the race," the universal mind. Hence the older faculty of imagination, which has direct access to it, is on the whole superior to the faculty of reason: "Trust to your own dream life" is the first principle of wisdom. Hence the poet has a "greater wisdom," remains our "best teacher," perhaps utters "transcendental truth." Prescott does qualify these statements; he has many wise things to say about the dream life, and at bottom recognizes the necessity of criticizing and controlling it. But his express desire to demonstrate the superiority of poetry leads him in effect to play down the rational faculty. When he asserts that "the oldest books are the best," because in olden times the imaginative faculty was stronger, the dream life was freer, he commits himself to an argument by which one could also prove that the savage is better than the civilized man, the child than the adult, and conclude that wisdom lies in letting nature rip.

This conclusion becomes explicit in Herbert Read's *Art and Society.* With the help of Freud's later concept of a superego—the level of self-observation, conscience, social ideals (in which, incidentally, I can see no essential difference in kind from the ego)—Read constructs a neat diagram of the esthetic process. Art "derives its energy, its irrationality and its mysterious power from the id, which is to be regarded as the source of what we usually call inspiration. It is given formal synthesis and unity by the ego; and finally it may be assimilated to those ideologies or spiritual aspirations which are the peculiar creation of the superego." The "primary function" of art, however, is "to materialize the instinctual life of the deepest levels of the mind." Here he parts company with Prescott. Like Freud, he assumes a "fundamental opposition" between id and superego; whereas the superego is responsible for the ideal values that for Prescott are the "greater wisdom" in poetry.* At the same time, Read has as high a regard for art. And so he logically calls for a transvaluation: an inflation of the instinctual values, a deflation of the intellectual and moral values. To be sure, he still wishes to "educate" the instincts (though by what standard he does not make clear); elsewhere he also says that any division of the total human personality is arbitrary; but actually he effects a sharp split, and he then takes his stand with the irrationalists. Look, he says, at the bloody mess that the superego has made of the world for two thousand years (as if it had been in control). Surely we have nothing to lose by giving the id more chance.

Well, the id is now getting its chance: Hitler is seeing to the deflation of the intellectual and moral values; and it seems that we have considerable to lose. Hitler also trusts to his dream life, which looks like nightmare. This is not, of course, the whole truth about him or the state of the world. But it is reason enough for being critical of the instinctual or the dream life, for taking Freud's own word about the unconscious mind. To consider this mind primary is only the more reason for emphasizing the need of commanding and controlling it. And this need is as plain in literature as in conduct. If the unconscious is the fount of inspiration, the artist still has no call as its high priest or prophet: scientists, philosophers, statesmen, engineers—all creative thinkers and doers have inspirations and high imaginings. All alike must consciously think and do, not only to realize their inspirations but to have them in the first place; imagination may be free, but it does not offer something for nothing. All alike must be forever critical. For the unconscious is also the source of blind fears and hatreds, all the blind impulses that enslave man; what comes up out of it is often trivial or absurd, when not morbid or monstrous. The artist cannot trust it even as a means to universality, a profounder humanity. If we all get together in Henry Miller's "brotherhood below the belt," we get together more distinctively and significantly in consciously shared meanings and goods. In short, conceive the unconscious as the matrix of

* An odd corollary of Read's thesis is his conception of form, which becomes a kind of sugar-coating for content. As the shapes of the instinctual life are "repellent," the artist consciously "invests his creation with superficial charms" —clarity, harmony, wholeness. Artists have often chafed at the discipline of artistry, as a violence done to their pure inspiration; but I wonder how many have ever thought of this inspiration as repellent, and form as a palliative.

all dream and desire, the source and the memory of our deepest experience, and it is still never to be cultivated at the *expense* of intelligence; for its whole force still has to be directed, its whole content evaluated, by the conscious mind for reasoned purposes.

This, to repeat, is Freud's own view of the matter. It does not imply an easy reconciliation. There remains the "deep place," in the words of Katherine Anne Porter, "where the mind does not go, where the blind monsters sleep and wake, war among themselves, and feed upon death." Even today men often forget the blind monsters. Psychologists overlook them as they talk of the unity of the mind, lovely integrated wholes; intellectuals in general overlook them as they hold up their rational ideal. And so, at the end, we do well to return to Freud's impressive contribution to knowledge. If the unconscious is not *the* psychic reality, and sexuality not *the* basic drive, they are still very important; his specific demonstrations remain largely valid. It is usually easy to translate his thought into other terms or adapt it to other purposes. He oversimplified causes, mistaking conditioned habits for unconditional instincts, but his discoveries about the event or effect can be set within a cultural frame of reference; he split up the mind into sharply opposed elements, but these can be brought back into organic relation. It is at any rate impossible not to use his thought. The notion of a divided personality struggling with itself, often unaware of its own motives, often seeking devious satisfactions—all the notions of repression, rationalization, projection, regression, compensation, sublimation have entered everyday habits of thought and speech.

For literary criticism, aside from the obvious biographical clues, psychoanalysis has illuminated the whole province of imagery and symbolism. It has not only discovered a wealth of unsuspected meanings but made clear their importance as an unconscious revelation of the artist's mind and character. In the depths of his imagery, Kenneth Burke remarks, an artist cannot lie; we can look forward to subtler, more acute studies of the kind Caroline Spurgeon made of Shakespeare. In general, Freud's whole analysis of dream-phantasy bears on the materials and methods of imaginative creation. The necessity of sensory images, the indifference to strict logic and literal realism, the substitutions and displacements, the remarkable condensation of material, the frequent ambivalence—all that he discovered in dreams elucidates the "wondrous agency" that Carlyle admired in symbols, their power of "concealment and yet revelation." And Freud's later theories allow a more adequate treatment as well of the motives of phantasy and art. In *Beyond the Pleasure Principle* he went beyond the bare wish fulfillment, as he had found that some dreams do not evade the unpleasant situation but go through it, seek to master it on its own ground —come to grips with it, that is to say, as tragedy does with the most fearful realities of human life.

Fundamentally, Freud's view of mental life is congenial to the artist, even apart from the ease with which it may be translated back into the ancient dramatic terms of the conflict between the powers of light and darkness, good and evil. As Lionel Trilling observes, Freud conceived the mind as in its primary activity "exactly a poetry-making organ." Likewise he illumined the deepest sources of poetry. "The primitive foundations of the human soul," writes Thomas Mann, ". . . are those profound time-sources where the myth has its home and shapes the primeval norms and forms of life. For the myth is the foundation of life; it is the timeless schema, the pious formula into which life flows when it reproduces its traits out of the unconscious." Here is the theme of Mann's *Joseph* story. "Life is quotation," an eternal anniversary; man does not exist alone in the here and now; and conscious celebration of the anniversary, identification with the myth, is the means to self-awareness and sanction. Thus Joseph relives the myth drawn from the unconscious, giving a "consummately artistic performance." His "smiling, childlike, and profound awareness" of his role is a celebration of the meeting of poetry and psychology, a symbol of the hope that "Where Id was, there shall Ego be."

Such literary translations of Freud's psychology may also emphasize its dubious scientific status, its need of stricter formulation. Meanwhile, however, they point to its value for the humanities. Freudian man is more interesting than Platonic man, or Rousseau's noble savage, or Mill's rational man, or Marx's economic man, or Nietzsche's superman, or Shaw's new Methuselah; and he still has a genuine dignity and force. He is always torn by conflict, threatened by the powers of darkness; his victories are compromises, invitations to further battle; yet he continues to aspire, he is worthy of the struggle, and his virtues emerge from it. Thus Mann believes that in Freud's thought lie "the seeds and elements of a new and coming sense of our humanity"—a humanism ironic yet not irreverent, and bolder, freer, blither, riper.

4. GESTALT PSYCHOLOGY

A MELODY is a related whole, something more than the notes that compose it; once heard, it can be recognized when played in any key. Von Ehrenfels observed that this simple fact of experience cannot be explained by a psychological theory of elements in mechanical association, for in a different key the elements, the separate notes, are physically and psychologically different. This observation was the origin of the theory of the Gestalt or "configuration," developed in the early 'twenties by Max Wertheimer, Kurt Koffka, and Wolfgang Koehler. Its premise is that the basic psychological unit is the whole, a whole that is not determined by its parts but that determines them. Gestalt psychology is an inquiry into the nature and behavior of such wholes. Although this organic principle had indeed been implicit in much psychological theory, from William James on, it had not been consistently followed out. The Gestalt psychologists made the principle explicit, demonstrated it experimentally, and applied it systematically to the phenomena of perception, learning, memory, emotion, personality—the entire realm of behavior and consciousness.

Thus innumerable experiments have shown that the "simple sensations" assumed or sought by analysts are in fact Gestalten, and that elementary relations do not have to be "learned" by experience. The field of vision naturally resolves into a dominant quality or "figure" against a subordinate "ground"; such perceptual properties as "round" or "symmetrical" are psychologically as concrete and immediate as red or green. These patterned wholes of perception will not necessarily be real physical wholes, as when we see a constellation of stars. Proximity, similarity, "uniform destiny," and various other factors determine the relations perceived. But in general the pattern tends to become as simple and orderly as possible with the given stimulus —to become the "good" Gestalt. We supply a missing letter in a familiar word and so overlook a misspelling, we usually see faces as symmetrical whereas actually they are not—from the beginning we literally put things into shape.* When we do realize that something is "wrong," as with a familiar room in which some change has been made, we often realize it before we see what is wrong; we perceive change as such. This selective principle is in general much like Wundt's principle of attention; both imply the active participation of the organism, which helps to determine what it takes in or makes out. Wundt, however, had to introduce a "faculty" of will to account for attention, just as all mechanists and elementalists have to invent some further principle to get things into relation again and keep them moving. The Gestalt psychologists hold that experience *comes* organized; they make organization, continuity, unity a principle of explanation instead of a problem to be explained. Nor is this an arbitrary verbal solution. Koehler asserts that consciousness takes after the nervous system, whose structural properties are also Gestalten.

Hence consciousness becomes less, not more mysterious. Similarly "intuition" becomes less mystical, and all "flashes of insight" less suggestive of divine inspiration. Thorndike and other psychologists have held that animals learn by haphazard trial and error: in thrashing around, the animal hits upon the correct solution to a problem, and by some process of mechanical association learns to repeat it. In *The Mentality of Apes* Koehler demonstrated that the animal has "insight." Without coaching, his experimental apes quickly learned to obtain a banana that was out of reach by knocking it down with a stick, pole-vaulting, piling up boxes as a ladder, and performing increasingly difficult stunts. Blind trial and error could scarcely account for even their simpler performances, but their behavior made it plain that they suddenly "sized up" the situation, grasped the proper relation of available means to ends. (Their errors, too, were often inspirational, like the wrong hunches of a scientist.) Hence the animal's ability to solve problems—call it intelligence, call it intuition—appears as a Gestalt phenome-

* Faces made absolutely regular by a photographic stunt—putting together two right or two left profiles—seem emptier, less interesting or lively than the normal face. Koehler suggests that the stress in forcing the perception of perfect symmetry may explain our greater pleasure in the slightly asymmetrical. This would also help to explain our preference for hand-made to machine-made laces, or for metrical irregularity to the singsong beat. We might say that to experience actively, "creatively," is to make good Gestalten.

non. In this it is similar in kind to human intelligence. Men can consciously analyze and abstract as well, but their knowledge begins as an intuition and ends as a formulation of relations.*

In personality or character the configuration, the "hanging-togetherness," is perhaps most conspicuous. A personality is never a mere sum of traits and cannot be explained by the most complete inventory. Yet in this field elementalism has been most rife and most naïve. Analysis has produced faculties, instincts, traits, motives, drives, and other units conceived like the parts of a machine or the ingredients of a pudding; statistical methods have produced such abstractions as "the average two-year-old child," and then encouraged the tendency to regard deviations from the average as eccentric or abnormal. Individuality has been either ignored in the classification of types and traits or regarded as the sum of incidental differences, like the blemishes that distinguish one apple from another. In *A Dynamic Theory of Personality,* accordingly, Kurt Lewin insists upon the need of beginning and ending with the total concrete situation, always in a full awareness that there is no such thing as an average man or an average situation, and that functionally considered, as Koehler remarked, a heart has more in common with a pair of lungs in the same organism than with other hearts. This total situation, moreover, is not merely the organism-as-a-whole but the organism in an environmental field. The Gestalten of behavior as of perception are the product of interaction.

At the same time, Lewin warns against the tendency to make wholes too inclusive or unity too perfect. He conceives the total personality as "a weak Gestalt comprising a number of strong Gestalten"; the psychologist's task is to study a given constellation of emotions, purposes, wishes, etc. If all are finally con-

* Koehler's principle has been confirmed by many independent investigations. E. L. Bouvier reported experiments showing that wasps, bees, cockroaches, and other instinct-ridden species can manage unexpected situations, overcome unnatural obstacles, and in general display the "discernment" that Fabre attributed to them. More striking are the observations reported by Jean Piaget in *The Language and Thought of the Child.* The development of the child's thought is a progress from the undifferentiated whole to the part; he has to learn to make out "elements"—to *dis*sociate rather than associate. Meanwhile he tries to justify everything and is unable to conceive of chance or accident, because in his world everything is connected. One might say that whether or not he has intimations of immortality, he is in this respect the "mighty philosopher" that Wordsworth hailed: he does not see life steadily but he does see it whole. The great minds have to make a great effort to recover what they were given in the first place.

nected, some connections are obviously negligible. My distaste for parsnips need not be correlated with my fondness for the novels of Joseph Conrad, nor the fact that I once had chickenpox with the fact that I am now discussing Gestalt psychology. Organized behavior would be impossible, indeed, if all processes were equally integrated with all others; organization implies differentiation, a *lack* of utter unity. If there is always some consistency or coherence, there is always some disorder or disharmony; and even the ideal of complete "integration" needs to be qualified. The adult can never be as perfectly integrated as an infant; the infant responds as a unit on the all-or-none principle, putting its whole body and soul into a howl. Growth is an integrated process, but it is also a process of differentiation, individuation—in a sense of *dis*integration. And so is consciousness, for in perfect oneness there would be nothing to be conscious of.

Now, even in theory the Gestalt psychologists have still not comprehended the whole man. Stern commented that they have neglected the man himself—the *Gestalter*. They have emphasized the whole field at a given moment; they have not treated satisfactorily the groundwork of individuality, the integrating drive or disposition that gives a life history its continuity or a personality its consistency and integrity. More objectionable, however, is their tendency to forget the arbitrary, merely convenient element in their solutions, and to believe that psychology can literally comprehend the whole man, literally represent the total situation. Hence they have often been unduly suspicious of any results got by analysis or any suggestions of mechanism, disregarding the real convenience of other methods and concepts. Hence Wertheimer talked freely of grasping the "concrete content" of intuition, personality, and other terms that hitherto have only named the problem, and he rejected as "defeatism" the idea that there are regions inaccessible to science. But if "intuition" is a name, so is "Gestalt," and if the concept of dynamic wholes is more fruitful than atomistic concepts, it is still something like a *deus ex machina;* in any event, whether or not there are regions inaccessible to science, scientific statements are necessarily limited by the logical requirements for such statements, and cannot represent completely the concrete content of experience. Actually, the Gestalt psychologists have been too quick to leave concrete experience. As other psychologists complain, they rush to sweeping philosophical generalizations. Ten years after the school had got under way, for example, Raymond H. Wheeler was confidently formulating the eight basic "laws of human nature."

Nevertheless these are incidental extravagances. Gestalt theory can readily accommodate the positive findings of other schools, for the wholes they study are structured wholes, the "parts" are still real. The mechanisms discovered by behaviorism, the sensations discovered by introspection, the drives discovered by dynamic psychology, the unconscious processes discovered by psychoanalysis may all be viewed as modes of behavior of the total organism, aspects of the total situation. Increasingly today they are so viewed. And here, as I see it, is the great importance of Gestalt psychology. It offers a fundamental unifying concept: a principle of continuity that holds alike for psychic events, physiological events, and physical events; a means of correlating the various branches of psychology with one another and all with other branches of science.

To begin with, although the Gestalt psychologists often appear more misty and metaphysical than other species, their basic logic is more strictly scientific. Lewin points out that most psychologists are still in bondage to Aristotle. Paradoxically, Aristotle was in a way more empirical than physicists are; he referred more directly to the actual and historical, measuring the "lawful" by frequency of occurrence. The propositions of physics are more purely conceptual: the laws of motion were established, not by getting an average of innumerable stones rolling down hills, but by conceiving the frictionless rolling of an ideal sphere upon a perfect plane. By the use of such fictions, physicists are able to apply their laws without exception to *all* phenomena. Psychologists, however, still roll stones, get statistical averages, state what happens "as a rule"; they regard the plain exception as unimportant or only a proof of the "law." Especially in the "higher" spheres of will and affect, they do not demand complete regularity or even know what to do with the fact of individuality, except smother it under more statistics or hand it over to "intuition." But in all spheres their empirical generalizations, useful and broadly true as they may be, are never scientific laws; in physics no stone can behave eccentrically, a single plain exception disproves the rule.

Gestalt psychology, however, will have nothing to do with averages, frequencies, exceptions, or with sharp distinctions between higher and lower, normal and abnormal. Lewin insists that it sticks to the assumption necessary for a science, that psychological events are as lawful and homogeneous as physical events. It investigates the particular situation in its "full concreteness" in order to discover laws that will hold without exception for all situations, comprehending at once optical illusions, insights, reflex adjustments, and neurotic maladjustments, just as the laws of motion hold for the wheeling stars and the falling leaf. It nevertheless enables the individual and the concrete to come into their own, since they must be accounted for as significant facts. It does full justice to the distinctive subject matter of psychology, neither attenuating it as introspectionists have nor impoverishing it as behaviorists have.*

Specifically, Gestalt psychology fits into the logical pattern of modern science as perhaps the most thorough, systematic formulation of the concept of organism. Even in the subatomic world, where physicists have sought to break matter down to its ultimate particles, they have found not merely order but organization, dynamic patterns of electrons. The facts dealt with by wave mechanics, says Planck, cannot be broken up into bits and the bits treated independently; "the picture must be held before the eyes as a whole." (Gestalt theory is also roughly analogous to the quantum theory in its principle of discontinuity; the patterned wholes of experience are fundamentally discrete, with "jumps" between them instead of a progression of intermediate patterns in a continuous scale.) In the macroscopic world, events still more clearly "take place" within a system; the theory of relativity, the principle that even time and space ultimately cannot be treated independently, makes modes of relationship the key to the whole cosmos. And in the world of living organisms, as we have seen, the organisms grow and behave as creative sys-

* A striking literary illustration of Lewin's principle here is the theory and method of Marcel Proust. He was aware that his whole past experience was "abnormal," not to say often trivial and tawdry; he was nevertheless confident that a complete realization of this—or any other—experience would have more universal significance than the thickest slice of life of the conventional realist. "People foolishly imagine," he wrote, "that the vast dimensions of social phenomena afford them an excellent opportunity to penetrate farther into the human soul; they ought, on the contrary, to realize that it is by plumbing the depths of a single personality that they might have a chance of understanding those phenomena." Everybody knows that Proust was an extraordinarily acute analyst. By Lewin's reasoning, he also turns out to be more scientific than most professional psychologists.

tems, by integrated whole processes. To give one more example, experiments on salamander embryos have shown that they at first swim awkwardly because of incomplete development, not lack of experience. Embryos that have been anesthetized, prevented from swimming during this awkward age, do as well as their practiced brethren once the drug is removed; at a given stage of maturity the whole system has "learned" perfectly by itself—just as water "learns" to keep an even surface the first time it is poured into a pan.

Hence the significance of Gestalt psychology is that it not only consistently applies the principle of patterned wholes to human behavior but locates it in our *phenomenal* world, the world we directly know, feel, live in. It makes clear that this principle is not a mere logical fiction; it links percept and concept, experience and knowledge. Theories of cognition, Dewey points out in his *Logic,* have alternated between a narrow empiricism, which explains the elements of sensation but not the "thought" that relates them, and a narrow rationalism, which accounts for the thought but not the elements; he maintains that relations as well as "things" are directly experienced. Of this epistemology Gestalt psychology is the concrete demonstration. Likewise it supports the doctrine of Bradley, that relations do not merely exist *between* things, like a rope, but imply an underlying unity, an inclusive whole. And it supports as well the common man's notion that he directly knows something about the world, is in touch with it. "Man never knows," Goethe said wisely, "how anthropomorphic he is"; he can never be certain that the wholes he perceives are actual—often, to repeat, they are not; but if a common principle governs both physical events and neural events, it is not surprising that he should correctly see actual wholes. Just how he does remains a mystery. Nevertheless it no more requires a mystical explanation than does the ability of apes to get hold of the banana or the ability of salamander embryos to swim. All "take place," all are "fitting." He who must have the ultimate can refer all these Gestalten to the Great Gestalter.

Here is also another way of easing philosophical worries about the problem of cause and determination. Causation applies strictly to parts, which are determined by the whole, but we can admit a principle of self-determination or "purpose" when we think from the whole to the part. Thus man, as a highly organized whole, can within limits direct and control his behavior; as he is also a part of larger wholes, natural and social, his behavior is also determined—as in practice, at least, he always recognizes. Similarly with matter and mind, permanence and change, homogeneity and diversity, the universal and the individual: in terms of the whole-part relation they are no longer antitheses—as in concrete experience they never were. Here is a means, generally, of not merely achieving logical synthesis but reclaiming phenomenological unity: the vital unity that the work of art makes us *feel.*

Hence Gestalt psychology is obviously rich in implications for literary criticism. In general, it is a congenial psychology for all art lovers because it restores to intellectual respectability our concrete experience, the immediate, naïve sensory impression that precedes logical analysis or verbal definition. Although philosophers and scientists are ultimately as dependent upon these impressions as are artists, they have traditionally regarded them as mere traps for truth seekers; now we learn on scientific authority that they may correspond with reality, or at least with the kind of reality conceived by scientists today. We may venture the pun that the artist, too, makes phenomenal discoveries. In particular, Gestalt theory is more pertinent to our immediate esthetic experience than is any other psychological theory. It helps to explain why we can apprehend a work of art before we "understand" it, much as an outfielder starts for the ball with the crack of the bat, and accept or reject before we know the reasons for doing so; like the artist himself we immediately perceive that something is or is not right, does or does not "belong." More important, this theory leads directly to the primary source of value in art: the immediate grasp of the organic quality of experience, the immediate *realization* of this quality—a knowing and feeling of the organism as a whole. Here is also the key to the richness of esthetic experience, the spread and pervasion of meaning. Sensuous qualities are naturally fused with other qualities of experience, so that even in immediate perception they have an emotional, finally a moral expressiveness: colors seem to be in themselves gay or somber, combinations of sounds solemn or frivolous, outlines—in nature or architecture—majestic or mean. "Sense qualities," writes Dewey, "are the carriers of meanings, not as vehicles carry goods but as a mother carries a baby when the baby is part of her own organism. Works of art, like words, are literally pregnant with meaning." *

In this view we can accordingly make out most plainly the dangers of esthetic analysis, which almost all critics recognize in theory but succumb to in practice. The analyst is always apt to break up a work into arbitrary "pieces" instead of natural parts. He is apt to forget that the work was not put together in a corresponding fashion, and that our final apprehension is of a unique whole, not a sum of parts. Above all, he is apt to give undue prominence to isolated elements and treat them as fixed quantities. Perhaps the chief reason critics so often misunderstand what an artist has tried to do, or condemn him for not doing something else, is that they regard certain abstracted "properties" —didactic, romantic, naturalistic, grotesque—as invariables, in themselves and under all circumstances objectionable.

Everybody knows the function of dissonances in music. Almost everybody will admit at least the possible or theoretical value of distortions in painting, gargoyles in architecture, comedy in tragedy. Yet nothing is more difficult in practice than a clear, full, steady awareness of the principle implied here: that all elements in art are variables whose value depends on their position in a configuration, their functional relation to the whole intention of the artist and the total effect of his work. In art,

* As I noted in an earlier work, Gestalt theory gives an illuminating perspective on the so-called impressionists in modern fiction—such novelists as Conrad, Proust, Lawrence, Gide, and Virginia Woolf. Realism in the nineteenth-century novel, even before it became avowedly "scientific" under Zola, was influenced by the scientific method of observation and analysis, and unconsciously committed to the assumptions of classical physics. The writer lived in a world of substantial things, things that led a thoroughly lawful existence in time and space and were always ready to have their pictures taken; he accordingly put his trust in an orderly notation of experience—what Proust called a "miserable listing of lines and surfaces." The impressionists, however, returned to the immediate, naïve sensation. Thus in *Mrs. Dalloway* Virginia Woolf stressed all the disorderly particulars, the discontinuous "quanta" of experience, that had been blurred by the generalizations of the realist. Conrad perpetually agonized over the problem of *rendering* every unique sensation, never merely pointing, naming, summarizing. Proust above all strove to realize the whole significance of the immediate impression, cut beneath the conventions that in the name of realism had obscured concrete reality. They all got different impressions, of course; they have their own obvious limitations; but in this perspective they are not merely eccentric or perverse. Although their method might even be called more genuinely "scientific" than Zola's, at least it is more genuinely artistic: the artist naturally seeks, not to present an abstract, but to realize the immediate quality, the actual feel of experience.

one plus one never equals two. A source of weakness in one writer may be a source of strength in another; one "fault" plus another may result in a far greater fault, or a lesser one; eliminate a plain fault and you may lose some value too. Unsympathetic critics can raise havoc with excessive writers like Dostoyevsky, Lawrence, and Proust by isolating and adding up their sins of excess; whereas the sins are inseparable from their peculiar greatness. Who can imagine a dispassionate Dostoyevsky, a reasonable Lawrence, a healthy Proust? Or a severely chaste romanticism, an exuberant classicism? Difficult questions remain. So far as I can see, Gestalt theory provides no criterion for asserting preference between different kinds of artistic wholes—a Gothic cathedral and a Greek temple, Wagner and Mozart, Pope and Shelley. Nevertheless it emphasizes the essential faults within any given style, and of all synthetic work: elements that do not "belong" because they are merely external, separable, mechanically added and not organically integrated. Altogether, it is hard to demonstrate just why the churrigueresque style—according to prevailing taste—is an inferior *kind* of style. It is not hard to understand why the total effect of the best Mexican churches is impressive, and why this "decorative delirium" may be less objectionable than the much less gaudy ornament stuck on many public buildings today.

Implicit in this principle is also the objection to music or painting that aims chiefly to tell or illustrate a story, and to poetry that aspires to be merely music or word painting. The natural possibilities of a medium are also necessary conditions; to ignore or defy them may result in an interesting tour de force, but a tour de force is a synthetic, not a natural whole. At the same time, Gestalt theory does not support the doctrine of absolute genres. The boundaries between the various arts are not so fixed as Lessing implied, for the senses are themselves organically related. Sensuous qualities spread and fuse; expressions like a "loud color," a "sour look," a "smooth voice" are an implicit recognition of the unity of the senses. "What is essential in the sensuous-perceptible," writes Erich von Hornbostel, "is not that which separates the senses from one another, but that which unites them; unites them among themselves; unites them with the entire . . . experience in ourselves"—and with all the external world that there is to be experienced. Because the sensuous is grasped only as form, moreover, we have here another key to the unity of form and content. "We must hear the 'so' of the parts," Von Hornbostel goes on, "the 'so' of their relation, the 'so' of the whole music—this is its form, and at the same time its content. You cannot have this content except in this form." Here is also the reason why in a profound sense the style is indeed the man: "What a man is I know by *what* he does and says; but still more surely and directly by *how* he does it and says it, and by *how* he looks." And here is the source of the knowingness in art. Our knowledge of something "is not hidden behind the appearance, but is beheld directly therein." Artists are accordingly right when they say that the essential in art is sensuous beauty—and also right when they say that it is not; for the perceptible is more than merely perceptible, the beauty is more than sensuous. "It is the same organizing principle," Von Hornbostel concludes, "which calls forth organism from mere substance, and which binds the stream of happening into wholes, which makes the line a melody that we can follow, and the melody a figure which we can see in one glance."

All human values, finally, may be conceived in Gestalt terms. In *The Place of Value in a World of Facts,* Koehler outlines a theory of "requiredness," a "something being demanded" beyond mere facts. Although this notion of requiredness does not yet have a definite position among scientific concepts, it is implicit in all scientific practice—in the very assumption that we should stick to the "genuine scientific method." Always some facts belong to a context and others do not. An "inhuman deed" is analogous to jarring notes, clashing colors, illogical arguments; all value-situations fall under Gestalt categories—the interrelation of subject and object in a context of requiredness. Koehler's development of this theory is a little disappointing, even though he claims no more than a modest beginning. He never applies it to a concrete problem of value; nor does he tackle the crucial issue of how to decide among conflicting notions of what is required—how to evaluate values. But at least he suggests a way of getting the good, the true, and the beautiful into relation. He demonstrates that ideas of value are not utterly alien to science, and that ethical and esthetic values can also have a logical place in a world of facts.

5. POSTSCRIPT: PSYCHOLOGY AND LITERATURE

MANY specialized studies in psychology offer leads that the literary critic might profitably follow up, such as the studies of Vernon and Allport in expressive movement, of Piaget in child psychology, of Allport, Stern, and others in personality, of innumerable men in abnormal psychology. For the purposes of conclusion, I shall touch upon only one of these—E. R. Jaensch's investigation of eidetic images.

Such images lie somewhere between after-images and memory images. Unlike the former they do not give the complementary color (the green that appears after one has been staring at a patch of red); unlike the latter they are visible, actually and plainly *seen*. Although it has long been known that some people have peculiar intermediate experiences, these sensations were regarded as pure self-deception. Jaensch has conducted innumerable experiments that prove them real. Often the images get clearer and richer as the subject concentrates; he will make out letters on a placard in the picture that had been flashed before him, letters that he had first not remembered or consciously seen. Always they tend to be more vivid as the picture or image is more interesting. Most children have eidetic images but lose them with experience and education; hence few adults have them—in most localities. But they are not symptoms of lower intelligence. Education explains their wide variation in frequency among individuals, localities, and races. And Jaensch shows that they decisively modify not only the individual's perceptual world but the world of his thought and knowledge—the whole structure of consciousness.

One implication of this study is that Gestalt theory needs some qualification. The perceptual world is built up and is not uniform; Jaensch points out that the dependence of sensory elements on relations or the whole configuration varies with different individuals and different stages of development. He therefore sees further proof here of the plasticity of human nature. Elementary impressions are themselves impressionable; most reactions do not follow immediately upon a stimulus but depend on the "mirror world" of consciousness, and the apparent tendency of evolution is to keep plastic and adaptable even this primary basis of behavior. Most important, however, are the implications that Jaensch draws for education and philosophy. The philosophy of our culture is rationalistic; it assumes that logic is the most natural and the deepest discipline, laying down the forms that reality must take before it can become an object of knowledge or "truth." Hence education has naturally taken logic as its model in all subjects. Jaensch does not deny the value and necessity of such training. But he calls for a recognition of the value and necessity of eidetic thinking, which has been ignored or suppressed though it is no less natural. Whereas the child is treated like an embryo logician, his mental structure during the eidetic phase of his development is closer to that of the artist. Indeed, the most productive logical thinking, as in science, does not proceed merely by the laws of logic but resembles the thinking of the artist and the child much more than logicians realize. Eidetic images can arouse and vivify all our mental powers in all subjects, bring about a closer integration of thought and perception.

Now it may be said that all this amounts simply to a technical restatement of the importance of imagination; and here is the point of this postscript. We often hear that psychologists have only rediscovered what writers have always known, only made jargon of poetry. Delete the "only" and this statement may be true enough. Certainly psychologists have as yet made no discovery that calls for a revolution in the creation or the criticism of literature. Such revolutionary movements as Surrealism at most point out only another incidental possibility of art; the Surrealists become silly when they proclaim that their way is the necessary way of the future. In general, psychological jargon is now so popular that we do well occasionally to translate it back into the old literary terms, in order to get perspective as well as a bit of style again. In particular, we do well to deflate the pretensions of applied psychologists. If some special scientific knowledge was required for the control of human beings, Pratt observes, Hitler would still be an obscure house painter.

Nevertheless psychology is much more than a modernistic version of the old story. Its important contributions are plain enough even on the grounds chosen by patriotic literary men. When the psychologist corroborates ancient intuitions he gives them a more precise statement and a more general application. He also makes them more substantial by separating them from all the dubious things that writers have intuitively known; if Bacon anticipated Freud by saying that the poet submits the shows of things to the desires of the mind, many other illustrious writers have divined that the poet contemplates the essence of things in complete freedom from desire. The fact is that since Freud we know a great deal more about the unconscious mind, the deep springs of dream and desire. So do we know more about sensation, emotion, learning, memory, habit formation—the conditions of consciousness and behavior. If psychological theories are as yet mostly rough empirical generalizations, they are still empirically useful. And they are always relevant to the interests of literary men.

Even the new hopes of a *science* of esthetics should not be simply laughed out of court, despite all the tests and graphs they have inspired. As D. W. Prall points out in his *Aesthetic Analysis,* sensuous qualities also have systematic relations, all works of art are relational structures. If we can make out no intelligible patterns in smells or tastes—no structural relations, for instance, between the smell of a rose and the smell of an onion—sounds, shapes, and colors do vary systematically, in a serial order. Prall illustrates by an analysis of temporal patterns in music and verse, showing, among other things, that "iambic pentameter" does not describe the actual rhythm in blank verse, and that the four-foot and three-foot lines in ballads have the same actual length. If this analysis does not take us very far into poetry, no science tries to tell everything; the demand for a complete explanation, Prall notes, is "the demand for esthetic experience itself instead of esthetic theory." Given the record of metaphysicians over the centuries, we should be able to content ourselves for the time being with the modest contributions of experimental and analytical esthetics.

More important, however, are the general suggestions for the practice of criticism. Psychologists call attention to the many ways in which a critic may go wrong, and indeed to the impossibility of his ever being utterly "right." Complete catholicity or absolute impartiality, Ellis Freeman points out, would be a denial of every psychological principle of habit, interest, need, attention, organization—of *function* itself. But this very awareness can enable a more enlightened interest, a more intelligent functioning. Furthermore, psychology encourages an empirical approach, attention to what artists themselves are actually doing. Most critics in the past considered Art in the abstract, deducing its eternal laws and essential forms from a priori principles; they dignified it chiefly as an instance of some other kind of truth, a means to some other kind of experience—moral, religious, philosophical. With all their reverence for it, they seldom paid it the elementary courtesy of accepting it tentatively on its own immediate terms. Its history, at any rate, suggests that the "eternal" element is not a specific form or content but the creative impulse and the needs it satisfies; not the abstracted "properties" of the objects produced by art but the nature of the experience induced by art. In other words, it suggests a psychological rather than a purely logical approach.

The striking example here, of course, is the work of I. A. Richards. In the first place, he makes clear the limits and the dangers of this psychological approach. Richards has turned out little specific criticism of poetry, considering the elaborate apparatus he sets up, but even in theory he has not given a circumstantial account of how poetry orders and controls all his "impulses," or how poetic stimuli differ from others. The present state of knowledge does not permit such an account. But a further reason for his failure really to produce the goods he talked about was a too narrow psychological theory that conceived poetry merely as events in the nervous system.* He presented the poet as one who has a "wonderful command and control" of experience yet who, as a poet, has no knowledge of it and is permitted to tell nothing about it; all the little impulses seem to obey the poet's magic wand impulsively, just for the fun of getting organized. Likewise his passion for staying inside the skin led Richards to neglect the whole social context of the experience. Poetry promotes finer adjustments, he declared; but like Freud he slid over the question of adjustment to *what*. Granted his ideal of fullness and freedom of life, the widest possible coordination with the least possible sacrifice of important impulses, the factors here are variables. The race, the milieu, the moment largely determine what is possible and important, and must influence the choice we usually have to make, in life as in literature, between width at some price of disorder and order at some price of narrowness.

* I am here referring primarily to his earlier work, especially *The Principles of Literary Criticism,* which made him perhaps the most influential critic of the time. In recent years he has considerably modified his principles, with a gain in breadth and balance—though also some loss in sharpness and suggestiveness.

Yet Richards had all the virtues of his defects. His concern with the actual processes of creation and appreciation enabled him to avoid the much more familiar excesses of criticism. He preserved the universality of art without recourse to eternal verities, locating it in the uniform impulses from which art springs, the profound similarities in purpose and need presupposed by the very differences in men. He united art and morality without tying them to any one code or creed; their relation is basic and organic, not mechanical or willed, in that the fine conduct recommended by moralists can be realized only by a fine ordering of responses. He recognized the peculiar esthetic meanings and values without keeping them pure or isolating them; art induces attitudes, has permanent aftereffects—makes a real difference to the whole mind by predisposing us to some kind of belief and behavior. He preserved, generally, all the essential values of art without making them ineffable, or transcendental, or ideological, or dependent upon any absolute sanction. Even before he began to broaden his approach, the main tendency of his thought was to bring art into vital relation with other major interests, and thereby to make for more freedom and fullness of life.

At least the work of Richards has proved unusually fertile. The conservative opposition has acknowledged its importance by giving it close attention; all the younger critics have sharpened their wits on it. He has encouraged others, notably Kenneth Burke and William Empson, to follow up leads from psychology. His *Practical Criticism,* with its experimental analysis of the "stock responses" that block an understanding of poetry, has launched a thousand studies. Altogether, he has demonstrated that if psychology is still a dubious science, displeasing to strict logicians and dangerous for lenient laymen, it is nevertheless a profitable study, and need not be considered a threat to poetry or to human dignity.

VII

THE SOCIAL SCIENCES

Man makes his own history, but he does not make it out of the whole cloth; he does not make it out of conditions chosen by himself, but out of such as he finds close at hand. The tradition of all past generations weighs like an alp upon the brain of the living. At the very time when men appear engaged in revolutionizing things and themselves, in bringing about what never was before, at such very epochs of revolutionary crisis do they anxiously conjure up into their service the spirits of the past, assume their names, their battle cries, their costumes, to enact a new historic scene in such time-honored disguise and with such borrowed language.

KARL MARX.

1. THE STATUS OF SOCIOLOGY

IDEALLY, sociology is the apex, the crowning fulfillment of the sciences. It will correlate their theoretical contributions and comprehend their actual consequences; it will extend the sovereignty of knowledge over the social life of man that provides the means and shapes the ends of all other inquiry. Actually, sociology is still a hodgepodge, a vast tangle and blur. It picks up the left overs of the other sciences; it sprawls over history, economics, political science, and anthropology; it spills over into education, government, and community hygiene. Like psychologists, social scientists have yet to establish a common base of operations. They have gathered many facts on their own, but they do not agree on criteria for interpreting the facts or directions for using them. They bring in their findings to journals and meet regularly in conventions, but these get-togethers are more social than scientific and emphasize how far apart they are in first principles of aim and method. And while they work at cross purposes to set their house in order, outsiders debate whether a house is there at all—whether sociology can ever be a science.

Consequently it is again well to begin by saying the worst. The peculiar difficulty of the social scientist is that he lives in the midst of his subject matter, yet can never get his hands on it, work on it in a laboratory, examine its structure; the physiology of a society is as intangible as its psychology. Although social experiments are going on all the time, he cannot perform strictly scientific experiments, capable of independent verification; no important social factor can be isolated, eliminated, or completely controlled. Although social processes are vast, intricate, and slow, extremely hard to observe, they also produce more fundamental change, more real novelty than is known in the other sciences; furthermore the social scientist is himself involved in this change, the product of what he is studying. And the natural result of these difficulties is that the issues are commonly oversimplified, with an unnecessary slamming of doors or triumphant forcing of open doors.

Thus Robert Lynd, in *Knowledge for What?,* contrasted the gravity of our social problems with the triviality of much social research. "Seismologists watching a volcano" he called the fact-finders, and he appealed to them to tackle manfully their main job, which is to reconstruct our whole society. Max Lerner enthusiastically applauded this "unashamed instrumentalism." He hailed Lynd for having staked out "the most spacious claims for the possibilities of social thinking"; he condemned the "detachment" and "objectivity" of other social scientists as mere "dodges to avoid thinking, devices for saving their skins." Both men are saying, at bottom, that now is the time for all good men to come to the aid of their country. Unhappily, these stirring words do not solve the actual problem (not to mention the question of just what a scientist *can* do about a volcano). "Social thinking" is not necessarily social science, nor a statement of pressing needs an adequate premise for inquiry, nor a high sense of civic responsibility an adequate technique. Actually, there is no dearth of social thinking, and ever since Comte claims have been spacious enough. Thinkers have been all too prone to take off easily from the is to the ought-to-be or the is-to-be—to offer as essential truth a great deal of high, wide, and handsome theory, often with a deal of unhandsome emotion. If the solid achievement of *Middletown* entitles Lynd himself to a political fling, the achievement nevertheless lay primarily in his objective analysis of an unreconstructed society.

Yet other social scientists have unquestionably been misguided by the desire to be scientifically safe and sound. We need more facts before we can safely generalize, they say; let us therefore put aside all theories until we have got more facts, each content to add his little bit until we have a solid foundation. There are too many Darwins in social science, Crane Brinton writes, before the Linnaeuses have got to work. The trouble, however, is that the facts are not simply there waiting to be gathered. Theory is implicit in the initial decision of what facts to look for, as well as in any notion of their significance. The foundation of a science is not a glutinous mass of little bits of knowledge but a broad agreement upon working principles, upon what we want to know and how it can be known. If there are indeed too many Darwins here, Galileo and Newton established the science of physics by formulating its basic principles *before* much experimental data had been accumulated; they told what a physical fact should be, gave the necessary directions for gathering and classifying such facts. Meanwhile the specific objection to many fact finders in social science is that they have taken over these directions as necessary for all sciences, fixed for all time. "There is a general agreement," Harry Elmer Barnes wrote approvingly, "that sociology can become a true science of society only in the degree to which it is able to appropriate and apply those exact methods of measurement and analysis which constitute the indispensable attributes of science in general." Because of this agreement, many have not derived their working principles from their own subject matter, asked themselves what we really want to know or ought to know about society. They have simply looked for facts that can be handled by the method of measurement, and then assumed that these facts must be significant. They have cultivated the illusion of exactness by a trivial or aimless counting and classifying. And because physics excludes the normative, they have shied away from any notion of value, purpose, end. In short, they not only distinguish social science from social action, they divorce them. They suspect all effort to make means and ends meet.

It would be comfortable to occupy a middle ground between these messianic and mechanistic extremes. But the problem is still to find a common ground, a solid basis for a sociology that will be genuinely scientific and also genuinely social. In *From the Physical to the Social Sciences,* Jacques Rueff goes behind the specific concepts to the logic of modern physics, the postulational, if-then method of thought. Cassius Keyser enthusiastically recommends his work as a concrete demonstration of how this method can be applied in social science and ethics. Unfortunately it is not concrete. Although Rueff gives an excellent account of the operation of such thinking, he seldom comes down from the realm of abstract theory; the few sample operations he gives for social science are chiefly in economics and chiefly statistical. He does not provide a comprehensive hypothesis, a criterion for selecting from the many hypotheses already available, a basis for sociology rather than more sociologies.

Similarly John Dewey's pragmatism is at this point not pragmatic enough. His basic argument is cogent, offering at least an excellent definition of the problem. Social scientists, he maintains, not only may but *must* eat the forbidden fruit of science, and introduce judgments of purpose and value. A social fact is a concretion of precisely the emotions, desires, ideas, and ideals that are systematically excluded from a physical fact, and it is unintelligible apart from purpose. (When a man jumps out of a high building, for example, the physical fact is simple enough: he falls, at a given rate of acceleration; the social fact is that he committed suicide, and the pertinent question is why he consciously did so.) For that matter, Dewey goes on, all scientists have definite purposes and policies; all experimentally interfere with given conditions in order to understand them better; all seek to control as well as classify, foresee as well as see. So should the social scientist interfere, experiment, strive for some control over events if only to understand them better. In any event his policy inevitably reflects social purpose and reacts upon it. To make mere description of the present order the whole aim, Morris Cohen adds in this spirit, is to imply an acceptance of this order and to slight the purposeful striving that is leading to a different order anyway. And Dewey's position seems the more reasonable because he insists, as always, upon a thoroughly experimental approach: the scientist must conceive social ends as hypotheses to be tested, never as fixed principles.

Indeed, a common reasonableness silvers everything; and all the difficulties are blurred by a definition of scientific method that makes it virtually synonymous with intelligent behavior. Dewey supplies no practical instruments to go with his instrumental formula. As Lyle H. Lanier points out, he suggests no specific hypotheses or program of action; he gives no concrete example of just how one goes about testing hypothetical ends or reviewing the consequences of a policy; he fails to say who will perform the evaluation and what will be the criteria of its validity. If he answers that a positive program must await further preliminary inquiry, he is betraying his own "method of intelligence": social decisions must be made continuously, and pragmatists can scarcely wait until all logical qualms have been allayed. Above all, although he insists on the possibility of experimental control in social science, his concern for individual freedom would almost certainly lead him to oppose anything like systematic control of behavior. Lanier notes that he has strongly opposed the program of Soviet Russia, the nearest approach to a comprehensive social experiment.

All this must seem very discouraging—at least to those who will accept as science nothing short of a complete formal system of verified truths, or as the immediate job of social science nothing short of the reconstruction of civilization. Nevertheless even the sociologist whose eyes are bright with purpose needs to begin with the given terms of his problem. Up to a point the present social order must be accepted, like all natural orders; it is a reality, which must be known before a better reality can be constructed or even rightly conceived; it cannot be measured and manipulated like the reality of physicists, defined and limited for convenience' sake. One can hardly hope that sociology will ever attain the exactness of physics, the measure of reliable prediction and effective control. Meanwhile disagreement is multiplied and exasperated by the natural disposition to come up with a single all-explanatory principle—the dream that inspires or deludes all scientists. Tarde found the key to the social process in imitation, Gumplowicz in war and conflict, Giddings in consciousness of kind; Marx discovered the famous "iron laws" that made history go all by itself and were bound to make his wishes come true; and today the ambitious sociologist is still apt to be less content than the physicist with merely partial or tentative explanations.

Yet I say the worst, again, simply out of respect for the supreme importance of sociology. Any study of man or by man must finally take into account that he is a social animal, and his mind a social product; no study is free from social influences. In this sense sociology is indeed the logical fulfillment of the other sciences, continuous with them all, comprehending them all. Whether or not it is a science, it has profoundly influenced contemporary thought, and whether or not it can be a pure science, it is a necessary branch of thought. No writer today can escape the influence or blink the necessity.

It is accordingly wise to give sociologists every chance to do their best for us, and not to brand them at once as pseudo-scientists. They have a reputable subject matter of their own, beyond the specialized studies—of the family, delinquency, criminality, etc.—that they have made to support their professional status. Human society may be considered a level of "emergence," but it is at any rate a definite type of natural organization. That it is impossible to discover general laws of society has never been conclusively demonstrated, least of all on scientific grounds; all thinking about social affairs necessarily presupposes that there are some such laws. If sociologists cannot be utterly objective or detached, they can at least observe an external course of events. If they cannot isolate simple causes and effects, they do find some kind of connection between social events, and they can hope to formulate invariant relations. If they cannot perform strictly scientific experiments, they can still observe the increasing number of actual experiments, as in education, government, and business. At the very least, they can employ an experimental logic. They can state hypotheses, making their premises explicit and then developing all that logically inheres in them—as Karl Marx did admirably; they can finally check these deductions by an empirical test—as Marx's system has been tested by the events, if not by his disciples. Even the standard of predictability, as Max Weber pointed out, is not so absolute as some scientists have supposed. Predictability is limited by the scientist's practical interest, as well as by the aspect of reality he is dealing with, and it is never possible in full concrete detail. Shatter a pane of glass, and no physicist can ever predict the exact size, shape, and position of every fragment—nor does he want to. In social affairs, on the other hand, we are actually able to predict a great deal with reasonable accuracy; social life is dependent upon just this ability. No more does the admission of irrational behavior and voluntary choice deny the possibility of general laws. Irrational behavior is not lawless; in principle it is closest to instinctive behavior, in effect it obviously permits significant generalization. As for the sense of freedom, we feel it most strongly when we are acting rationally, and rational action is highly predictable.

Such arguments, furthermore, are buttressed by solid gains. All the controversy over premature or too spacious claims has obscured the actual progress toward a science of society. Apart from the miscellaneous factual findings, many important ideas survive this controversy; Comte's demonstration that the division of labor has wide cultural as well as economic consequences is but one example. All the leading sociologists have made some positive contribution that is now taken for granted. More important, there is now a pretty general agreement upon some basic concepts. To appreciate the advance, one has only to recall the social "contract" or "covenant" theories that once governed social thinking: the assumption of Hobbes, Locke, Rousseau, and others that men got together more or less deliberately in order to pool their resources, or to restrain their natural viciousness, or somehow to promote their welfare; the assumption, generally, that institutions are rational and willed, and that social causes are to be sought in conscious purposes. Today almost all sociologists, I suppose, would accept the proposition that society is not a mere aggregation of individuals but a real organism, and that the individual is inseparable from it. If this idea goes back as far as Plato's *Republic,* sociologists were the first to work out its implications systematically. Likewise they take literally, as a premise of inquiry, the old idea that morals are mores, historical and institutional, and discard the as old idea that they are the product of some kind of disembodied conscience. Sumner's detailed study of "folkways" is the obvious example of the conception that historically explains and logically justifies the rise of sociology—the conception of the far-reaching, all-pervading influence of the social environment.

Altogether, sociologists agree that societies have not been planned but have grown like organisms; they study the tremendous drama pictured by Engels, in which there emerges, from the coöperation and conflict of individual wills, something willed by no one and foreseen—if at all—by but a few; they study the deep, intricate processes that give a society its form and pressure, condition the individual's will from the outset. Beneath all controversy they are united in their basic concepts of the "dynamics" of the social "organism," their search for the principles of social "evolution" and "integration"; and these concepts link their inquiry with the other sciences.

Upon a closer examination, indeed, the convergence of leading sociologists becomes even more impressive. For they are also taking into account motives, values, norms, ends—the whole "social equation" that fundamentalists in science have con-

poseful elements as precisely the most meaningful, the index of the peculiar *social* reality, and they are introducing new concepts in order to handle them. At the same time, they make the necessary distinction between recognizing actual ends and determining proper ends. Their immediate purpose is to understand society, not to reconstruct it.

Perhaps the boldest statement of this position is Karl Mannheim's *Ideology and Utopia.* Mannheim shows that a mechanistic scheme, restricted to external, measurable relations, is the more inadequate because it cannot handle any *situation,* the living context of all social problems; the organic concept is the more valuable because it enables the scientist, like the common man, to "grasp a situation." But thought itself, he goes on, is inseparable from the situation, the living context. Even after it has been purified of its obvious wishfulness, it contains an "activist" element that is seldom explicit and can never be eliminated; purpose and action are not added to perception but are *in* it. As Napoleon said, *"On s'engage, puis on voit."* Hence there cannot be a purely objective sociology. The sociologist's first duty is to systematize doubt, to penetrate to the irrational foundation of all rational knowledge, and specifically to look into the social origins of all modes of thought. Mannheim accordingly analyzes different modes and relates them to the groups from which they spring. He distinguishes two main types: "ideologies," which seek to maintain the *status quo,* and "utopias," which seek to change it. At the same time, he does not consider this fundamental bias of all thinking a mere liability. In general, it is what enables us to see, gives thought its real use; without the utopian mentality, for instance, "man would lose his will to shape history and therewith his ability to understand it." In particular, any given bias enables one to see important aspects of the situation that others overlook. Nor does all this imply a complete denial of objectivity. It is a perception of real and important relationships, hitherto obscured, and it leads to a new conception of objectivity that makes possible more real freedom from bias. Just as self-analysis is the means to self-knowledge and self-control, so can sociologists bring the conditions of their problem within the range of objective knowledge and possible control. Thought becomes more rational as it becomes aware of its function, its methods, its relation to the whole life process.

sidered merely a source of error. They are regarding these pur-

This "sociology of knowledge" is an elaboration of the familiar idea of cultural compulsions or climate of opinion. It is not, as Mannheim leaves it, a whole science or even an adequate base for scientific operations. He formulates some important principles but not a complete scheme, some useful methods but not a complete methodology; he makes some suggestive analyses of ideologies and utopias but establishes no criterion for determining the validity or relative objectivity of such analyses. Nevertheless he does grasp some significant relationships, and he points to a way of comprehending the various systems that pretend to be comprehensive. He includes all of Marx's thought, for example, but also dates and places it as a utopian mode. He dates and places his own scheme as well. The stable society of the Middle Ages had a unified world view, which was appropriately absolutist; the very disintegration of the modern world has enabled us to perceive the relativity of values and points of view, the social origins and uses of thought. This conception is neither a perverse invention nor a brilliant discovery but a natural outcome of social change.*

In a recent work, *Man and Society in an Age of Reconstruction,* Mannheim concretely applies these principles to argue the necessity of a planned society. Our complex civilization demands "planned" or "interdependent" thinking—the organic mode of thought that it has given rise to. One thing does *not* lead to another, in such simple succession as Adam Smith conceived. (Technical progress raises profits, profits bring increased demands for labor and therefore increased wages, increased wages raise the birth rate and the supply of labor, the increased supply enables further division of labor and therefore more technical progress—and then the cycle starts all over again, ever onward and upward.) Rather, one thing leads simultaneously to many other things, just as many other things have led to it. Only organic thinking can deal with the multiple possibilities of this

* Our woes are thus the forge of our weapons—and the obvious moral is to make full use of these new weapons. In this view, absolutism is the cardinal sin of thought today because it not only fails to achieve an actual integration, having no social basis, but surrenders our positive gains, obscures the real meaning of the present. If men do succeed in setting up a thoroughly stable society, recovering a thoroughly unified world view, they will most probably become absolutists again; relativism would then be an anachronism. Meanwhile the terms of thought and social life are rather different. The totalitarian states have to suppress independent thought and depend on incessant, high-powered, centralized propaganda—the simplest and most superficial form of integration,

many-dimensioned reality, only a planned society can efficiently operate its intricate machinery. In this work Mannheim has accordingly answered Robert Lynd's call. But he continues to maintain a genuinely experimental, scientific attitude. He is aware, for instance, that there is no easy, guaranteed answer to the crucial question: Who is to plan the planners?

At any rate, Mannheim makes out a strong case for the "sociology of knowledge" as a necessary supplement to logic. Philosophers and scientists alike, he declares, have been concerned too exclusively with their own kind of thinking. They have neglected the kind men use in practical decisions, thought that cannot be separated from vital impulses and from the situation it deals with; they have considered impure the kind of knowledge got from experience, forgetting that we learn really to know other men only by living and acting with them. In this way science too has tended to divorce theory from action, thought from the situation. A definite type of mentality has plainly influenced the form it has assumed, so that most men take for granted that scientific thought is utterly different from ordinary thought, or even alien to it. Altogether, Mannheim conclusively proves the importance of sociology, he clarifies its possibilities as well as its difficulties, and his work is pertinent at once for positivists who are unwilling to admit sociology into the family of sciences and for sociologists who are too willing to curry favor by putting on positivistic airs.

More impressive for severe logicians, however, is the "voluntaristic theory of action" that Talcott Parsons formulates in *The Structure of Social Action.* It is more systematic and comprehensive, and it is more significant because Parsons derived it from a close analysis of four apparently very diverse thinkers: Marshall, Durkheim, Pareto, and Max Weber. The positivist-utilitarian sociologists, such as Comte and Spencer, had either eliminated social ends and norms or admitted them only as random wants—the given data, like the biological needs of animals, on which the sociologist erected his universally valid system. They assumed that their facts were of the same order as physical facts, and that the social reality was the sum of all facts; Spencer therefore regarded sociology as the encyclopedic science, not as a branch with its own subject matter. The voluntaristic theory of action, however, assumes that common values and ends—common to any given society but differing in different societies—are the peculiar "social factor," and that they are not simply given, not like physical facts. At the same time, they are not simply arbitrary or purely "spiritual." Idealists had recognized the importance of values in social action but had referred them to something immaterial or transcendent; in the voluntaristic theory they are natural "emergents," which interact with material conditions in determinate ways. Parsons maintains that the fundamental emergent property of "common-value integration" is not a mere fiction but refers to something observable, real. Sociology should accordingly aim to develop "an analytical theory of social action systems" in terms of this property. And the convergence of these four thinkers upon this new conception is the more impressive because they traveled independently and from markedly different points of departure. Marshall set out from classical economics, Pareto from classical mechanics, Durkheim from a radical humanitarianism (a favorite butt of Pareto's satire), Weber from German idealism. None was aware of their common direction.

Weber's *Protestant Ethic and the Spirit of Capitalism* is perhaps the clearest illustration of this theory. Although Weber did not deny the great importance of material factors in social change —all that Marx analyzed—he demonstrated that these factors alone cannot explain change. Specifically, they do not explain the *genesis* of capitalism. Thus capitalism did not develop at relevant stages in the culture of China, India, and Judea, even though material conditions were favorable to its development; it did not develop so fully in the Southern states as in New England, even though the colonies in the former were founded by large capitalists for business reasons. For it is also a spirit or *ethos,* a set of values and value attitudes, and the Protestant ethic was as much its cause as its effect. Because there are diverse value systems Weber believed, like Mannheim, that there could be no one universally valid system of sociology. Nevertheless he believed that the number of such systems was definitely limited, and that once the value element had been admitted it could be studied objectively; this study was not to be confused with value judgment. In short, Weber conceived a synthesis of science and social action. As scientific investigation is also a mode of action, which can be analyzed like any other, a knowledge of action is indispensable. But the verifiable knowledge of science is in turn indispensable for this very analysis of action.

This synthesis corresponds to Dewey's conception of the logic of inquiry and to the recent developments in scientific thought, which has become self-conscious and social-conscious. For the specific purposes of sociology, meanwhile, Parsons maintains that the voluntaristic theory is a genuine scientific hypothesis. It is supported by great masses of facts; it has been carefully tested and not yet refuted; it is useful as a basis for further inquiry. It will be modified and may be superseded, but he is confident that like the theories of classical physics it will leave a permanently valid residue, a truth that will be incorporated in the broader system. In this basic sense it is valid now.

For the layman, all this may be highly theoretical. He is likely to remain more impressed by the obvious differences among sociologists—and by the difference between all their theories and the positively verified or verifiable theories of physics. In fact, he ought to keep these differences always in mind; sociologists are not solidly united, and what many have in common is still a disposition to regard hypotheses as necessary truths. Yet all this can help one to find his way around; and it is well for literary men to get around. There can be no thorough understanding of literature without reference to social action, no comprehensive evaluation of it without reference to social ends; meanwhile sociological theory has already had a marked influence upon creative work as well as criticism. For such reasons, at any rate, I propose to examine the systems of Marx and Pareto, the one because it has been the most influential, the other because it offers an instructive contrast. They are not only especially pertinent for the literary critic but representative of the main tendencies in sociological theory, the fundamental agreements and also the chief sources of disagreement. They help one to place other systems, all of which could not be treated adequately in a chapter.

2. MARX AND MARXISTS

"Few literary dishes," wrote Thomas Huxley, "are less appetizing than cold controversy." Right now it is a little depressing to return to the ground where the flower of our advanced thinkers once fought stirring battles over the theories of Karl Marx, and where today a few ghostly survivors dispute the right to praise or to bury him. Nevertheless his system remains perhaps the most impressive as well as the most influential achievement in the field. It has stood the test of time better than most and was well built to stand this test: like a good scientific hypothesis, it provides in its premises for future development, revision through self-criticism and adaptation to inevitable social change. The good that Marx has done has also lived after him, in the thought even of those who attack or ignore him. And his failures are more instructive than the successes of most thinkers, especially for one interested in the possibilities of social science.

The most necessary thing to understand, if only to do full justice to Marx, is that his system is strictly not a sociology, not a science even by generous definition. It is essentially a political program. That Marx considered it scientific is simply the key to his fundamental inconsistencies. Now he was a dispassionate analyst of capitalistic society, picturing the nonmoral, inevitable historical processes; now he was an impassioned evangelist, exhorting the workers to put down the hideous crime of capitalism. (Not to mention the lonely, pain-racked, embittered exile who wrote to Engels, "I hope that the bourgeois as long as they live will have cause to remember my carbuncles.") At first he attacked social legislation like the ten-hour factory law as reaction in disguise, an idle effort to defy economic law and put off the foreordained revolution; later he rejoiced in such political victories, though they still broke his law of the inevitable increasing misery of the proletariat. But from first to last his primary interest was the Revolution, and to it he sacrificed, piously but unconsciously, the first principles of scientific inquiry. He set up his hypotheses as essential truths, he translated their logical requirements into inalterable necessities of nature, he asserted that the solution of his problem was given in advance (an assertion that in the exact sciences would brand a man as a quack) —he ended, like Martin Luther, in an absolutism as despotic as he had begun by demolishing. And if Marx accordingly needs to be rescued from his own excesses, he needs still more to be rescued from Marxists. Instead of subjecting his program to the constant criticism that dialectical materialism itself calls for, his followers have made it still more like a religion and less like a science, and chiefly wrangled over the proper interpretation of the holy text. He rationalized a myth; they mythologized his rationalization.

Hence the very familiarity of Marxism makes it advisable to start all over again, with the fundamental philosophy. Engels summarized it in *Ludwig Feuerbach:*

The great basic thought that the world is not to be comprehended as a complex of ready-made *things,* but as a complex of *processes,* in which the things apparently stable no less than their mind-images in our heads, the concepts, go through an uninterrupted change of coming into being and passing away, in which, in spite of all seeming accidents and of all temporary retrogression, a progressive development asserts itself in the end—this great fundamental thought has, especially since the time of Hegel, so thoroughly permeated ordinary consciousness that in this generality it is scarcely ever contradicted. If . . . investigation always proceeds from this standpoint, the demand for final solutions and eternal truths ceases once for all; one is always conscious of the necessary limitation of all acquired knowledge, of the fact that it is conditioned by the circumstances in which it was acquired. On the other hand, one no longer permits oneself to be imposed upon by the antitheses, insuperable for the still common old metaphysics, between true and false, good and bad, identical and different, necessary and accidental. One knows that these antitheses have only a relative validity; that that which is recognized now as true has also its latent false side which will later manifest itself, just as that which is now regarded as false has also its true side by virtue of which it could previously have been regarded as true.

Here is the ideal statement of the Marxist philosophy—and the measure of its shortcomings in practice.

Marx did begin with this basic thought, which is pretty much the standpoint of modern science. He also borrowed from Hegel the concept of the transformation, not merely the conservation of energy, the "qualitative leap" at a certain point in quantitative changes, which foreshadows the theory of emergent evolution. His dialectical materialism was accordingly a higher synthesis of mechanical materialism, which had neglected the active powers of man, and idealism, which had neglected the material conditions of life; he recognized the qualitative differences because of which mind is not identical with inorganic matter. He then began to woo the absolute with his dialectical "laws"—the mutual penetration of opposites, the negation of negation, culminating in a new Trinity of thesis-antithesis-synthesis—which he regarded as inviolable commandments of nature, instead of convenient descriptions of it—or, in human affairs, an elaborate way of saying "compromise." Nevertheless his dialectical materialism did make provision for new developments and hence for necessary readjustments. "With each epoch-making discovery in the department of natural science," Engels wrote, "materialism has been obliged to change its form."

Now since Engels wrote, there has been an epoch-making development in physics, a test to order for Marxists; and their official response to it was significant. I take for my text Lenin's *Materialism and Empirio-Criticism.* Although Lenin perceived that matter had "disappeared" only in the interest of a more comprehensive concept of energy, and that its disappearance did not destroy the existence of the outside world or the validity of our knowledge of it, he finally missed the point of the revolution in physics. He missed it because his object was *not* to adapt dialectical materialism to the new knowledge. His book is a ferocious attack on other Marxists—Bogdanov, Bazarov, Lunacharsky—who were earnestly making this effort, trying to be genuinely dialectical, striving for a still higher synthesis. For his almost pathological fear of the word "idealism" (which like "romanticism" to Irving Babbitt became a label for everything he disliked) drove him to a naïve, hidebound materialism that would accept no criticism even of its form. This made him simply incapable of understanding the logic of science; and he became the intellectual as well as the political leader of the party.

Specifically, Lenin refused to admit the elementary distinction between questioning the existence of the external world and questioning its complete knowability. He refused to be content with the pragmatic assumption that somehow we can and do make of the mysterious thing-in-itself an available thing-for-us. "There is absolutely no difference between the phenomenon and the thing in itself, and there can be none," he insisted; and despite this very bold assertion, which no scientist would ever attempt to prove, he bristled at any suggestion that materialism had a "metaphysical" element. He therefore objected furiously to Bogdanov's use of "energy" as a symbol, to Mach's "principle of economy of thought," and to all such working principles that are now commonplaces of scientific thought. Any suggestion that scientific statements were "prescriptions for practice" instead of exact reproductions of reality he denounced as a "capitulation . . . of science to theism." Any tampering with his solid, three-dimensional lumps was a betrayal of the proletariat. Actually Lenin, not Bogdanov, was the reactionary here —and the more profoundly because his materialism did not leave room even for dialectics. With him it was all thesis and antithesis; you were either an orthodox materialist or a reactionary idealist; and if you attempted to "reconcile" or "transcend"—to achieve synthesis—you were the more contemptible because a hypocritical idealist.

Here is the tyrannical, thoroughly unscientific attitude behind all the specific limitations of Marxist theory. It appears incidentally in Lenin's ugly controversial manners: he savagely abuses the motives as well as the mentality of his deluded fellow-Marxists, he repeatedly calls them "cowardly" and "traitorous," he describes their ideas as not only false but "foul"—he sets the key for the violent factionalism that has made a purge the logical way of settling an argument. It appears more significantly in his constant appeals to the text of Marx and Engels; the philosophical ideas of his opponents are not debatable or even merely illogical but "incorrect," "reactionary." It is explicitly stated as gospel: "You cannot eliminate even one basic assumption, one substantial part of this philosophy of Marxism without abandoning objective truth, without falling into the arms of the bourgeois-reactionary falsehood." Even as Lenin wrote these words, exact scientists were drastically altering their own basic assumptions; but he saw only a case of labor pains: "Physics is giving birth to dialectical materialism." The "incorrect" ideas prevalent among physicists were easily explained: "Physicists lack knowledge of dialectical materialism." Briefly, with Lenin dialectical materialism came first. In theory, the philosophy of Marx derives its authority from science; in effect, science derives authority from it.

Hence Lenin might not have been troubled had he realized that science has never made use of the Marxian dialectic. Liberally interpreted, its laws can be adjusted to the new concepts of science, can find room for the compromise he scorned. Nevertheless they are not scientific laws, and indeed are too cloudy to serve even as scientific hypotheses. Max Eastman has observed that the references to the dialectic are always vague or fragmentary, implying a large body of knowledge—which simply does not exist; Marxists have never found time to ground their logic. At best, it has the limited usefulness of Mannheim's "utopian" mode of thought. Trotsky acknowledged as much when he declared that "the will to revolutionary activity is indispensable to understanding the Marxian dialectic." This means, in Eastman's plain English, "In order to perceive with accuracy, we must conceive with prejudice." This is to repeat that Marxism is essentially a political program. But it is also to repeat that the program is not scientific. Ultimately, the materialism of Lenin, as of Marx before him, is grounded on the metaphysical moonshine of German idealism. Like Hegel, Marx used the dialectic to identify the Rational and the Real, to guarantee his ideal Reality. And so he deserted the "great basic thought," was himself "imposed upon by the antitheses . . . between true and false, good and bad," and forgot the "relative validity" of his own truth. He fell victim to his symbols and abstractions—the logical fictions that all honest-to-God materialists, Lenin insisted, must accept as literal copies of reality.

Against this philosophical background, the specifically social theories of Marx stand out more clearly. To begin, his important contributions derive more from his materialism than from his dialectic: the basic assumptions that all forms of culture are rooted in the material conditions of life, and that the consciousness of men does not determine their existence but their social existence determines their consciousness. He has forced a general recognition of the fundamental importance of economic life, showing conclusively that the Useful also conditions the Good, the True, and the Beautiful. Moreover, a great deal of his specific analysis of capitalism—its genesis, its workings, its internal contradictions—has an objective validity that is denied only by the economic illiterates in legislatures, corporations, and some editorial offices. *Das Kapital* remains the most comprehensive, illuminating study yet made of capitalism, its obvious triumphs and failures as an economy and also its effects on culture: how the constant revolutionizing of the means of production has led to as constant and drastic change in social relationships, endless uncertainty and agitation, activity the more furious as its goal becomes less clear; how old pieties have been destroyed, cultural values translated into cash values, private morals cut off from business practice; how class interest has affected government, law, religion, philosophy, art, and all other cultural institutions. No thinker can have a sound notion of how and why our society became as it is, and what might be done about it, without acknowledging a heavy debt to Marx.

Perhaps as important as the analysis, moreover—or even destined to outlive it, as Plato's idealism has outlived his ideas—is Marx's ideal of the classless society. Engels hailed the proletariat's destiny of "once for all, emancipating society at large from all exploitation, oppression, class distinction, and class struggle." This is a rather visionary conclusion for a scientific demonstration, an illogical happy ending for the dialectical processes of history that imply constant struggle and no ending. Nevertheless the classless society may be regarded at least as a reasonable hypothesis for social experiment. It is the more reasonable because some form of socialism is a logical correlate of vast industrial organization and increasing social interdependency. In any event, Marx's thought has been a positive social force, and the force has derived from the ideal as well as the logic, the classless society as the Good Society and not merely the inevitable outcome of historical processes. Men do not fight and die to establish a foreordained fact. And this ideal value needs to be stressed, partly because it has been overshadowed by the harsh elements of his creed, partly because it has been played down by Marxists themselves, who are likely to be embarrassed by such language or even contemptuous of it.*

Many humanists have accordingly been attracted to Marx's philosophy. In theory it leaves room for their values; in practice it is designed to incorporate these values in the form of society. At least the humanists should accept Engel's doctrine that freedom is the recognition of necessity and social consciousness the means to this freedom, the author of our "better self." Meanwhile the often intolerable compulsions of social consciousness result from the glaring inequalities and injustices of capitalistic society; and these Marx has exposed more effectively than any other thinker. Though he was often duped by his beloved abstractions, he did return repeatedly to the concrete human re-

* Thus Lenin attacked Lunacharsky for speaking of "religious atheism," "scientific socialism in its religious significance," its "deification of the highest human potentialities." But communism plainly is the equivalent of a religion, complete with scripture, theology, gods (including Lenin himself), and even, Kenneth Burke notes, the Christ-like metaphor of the suffering proletariat that is to redeem the world. The soundest argument for it as a social program is that it actually can excite religious emotion, ideally may realize the highest human potentialities. If leading Marxists had seen this more clearly, they might have succeeded better in distinguishing religious fervor from scientific analysis, and themselves been less often taken in by the intellectual indignities and absurdities that inevitably accompany a popular religion.

lations behind them; though his logic is often unconvincing, his documentation is always compelling—a terrific indictment that cannot be quashed by mere logic.

All this is to say a great deal—and at that perhaps not enough. The final proof of Marx's importance, that so much of his thought is now taken for granted, makes it difficult to do him full justice. Yet it is also proof of his importance that his most dubious dogmas have had as wide influence. Today it is the more necessary to separate these dogmas from his positive contributions, so that they will not be thrown overboard together.

The main source of the now evident faults of analysis—the iron laws that have been broken, the inevitable outcomes that have not come out—is Marx's gross simplification of a complex. He tried to reduce intricate interdependencies to a single relation of cause and effect, to force organic processes into a one-way traffic. Specifically, the basic weakness of his system is the doctrine of economic determinism that he made its shibboleth. He qualified it, to be sure; the economic factor was supposedly the ultimate determinant, not the only one. Nevertheless he greatly overstressed this factor (as Engels admitted in later years) and neglected the other basic needs that motivate man's doing and making. In practice, his economic determinism was a rigid economico-mechanics, after the model of classical physics. It is accordingly inadequate because it fails to take into account precisely the most distinctive characteristics of any social organism. Marx brilliantly analyzed the action of economic forces, their influence upon all other social forces; he failed to analyze the *reaction* of these other forces, the *interaction* of the whole complex. He did not consider how cultural forms in turn modify the material conditions of life, how consciousness in turn affects social existence.

Marx and his followers have therefore been most successful in predicting the course of economic development—the growth of technology, the increasing centralization of capital, the intensified struggle for markets, the closer alliance between government and industry, etc. But they have failed pretty badly in foreseeing the larger social consequences of this development. Perhaps most conspicuous is their failure to foresee the political uses of modern technology—the tremendous power of the German military and propaganda machines—and to face the whole problem of political power. The state was destined to "wither away," simply because they had defined it away. Actually, power must be exercised by a minority, if not by a leader; the problem is to assure that it will be exercised intelligently, in the interests of the whole people; and Marxists have solved this problem in theory by denying its existence, in practice by excommunicating or liquidating anyone who questioned whether Stalin was always and automatically expressing the "real" or "true" interests of the workers. Actually, the state in a highly industrialized society does not shrink but expands, takes over more and more administrative duties, as it has alike in Russia, Germany, and the United States. Max Weber was far sounder in stressing the continuity rather than the differences between capitalism and socialism, for socialism only accentuates this central fact of bureaucracy.

Similar Weber's analysis of the *ethos* of capitalism points to another cluster of social forces that Marx did not take into account. He neglected the active power of ideas and ideals, which has made possible the history of Marxism itself. He neglected the power of knowledge, the measure of objective thought that enabled him, a bourgeois intellectual, to discover the cause of the proletariat and map out a campaign. (In a thorough-going economic determinism Marxists would even be deprived of enemies, for there could be no reactionaries; to be behind the times would be as impossible as to be ahead of them.) He neglected the creative force of the genius, the heroic personality, the great leader: As with Alexander, Caesar, and Napoleon, so with Lenin; for without his genius—and his courage to defy the law that the proletarian dictatorship must be established in the most advanced capitalist state—there might have been no Russian Revolution. All such "ideal" powers are incalculable, of course, and introduce an untidiness displeasing to scientist, philosopher, and reformer alike; we have as yet no certain way to measure their force, foresee their direction, control their effects. To ignore them is nevertheless to ignore the given conditions of the problem. And Marx's wishful faith in impersonal, automatic processes was also profoundly inconsistent with his emphasis upon knowledge for control. A consciousness of social forces that have been operating blindly and destructively, like a knowledge of the forces in nature, brings a possibility of control, of an approach to the ideal state predicted by Engels: "humanity's leap from the realm of necessity into the realm of freedom," where "men, with full consciousness, will fashion their own history." Ironically, Marx has probably contributed more than any other thinker to the infraction of his own laws, for he has strengthened the countertendencies that he neglected, the tendencies in the advanced capitalistic countries toward amelioration of the class conflict, compromise and readjustment short of revolution, experiments in social planning, efforts at social control.

From all this it follows that the primary equations got by the formula of economic determinism—the interpretation of history in terms of the class struggle, of modern society in terms of the struggle between proletariat and bourgeoisie—are also inadequate. Class conflict obviously explains a great deal in history, but as obviously does not explain many events of immense consequence, such as religious wars, barbarian invasions, the conquests of Alexander, the spread and the later revival of Greek culture, the invention of printing and of gunpowder. As for the modern world, Marx's proletarian-bourgeois antithesis is perhaps his grossest abuse of symbols and abstractions. He overlooked the identities in interest and desire that unite boilermaker with bookkeeper, cause them to resist the most earnest efforts of intellectuals to make them class-conscious, justify the politician's appeal to the great American people. He slighted as well all the divergences within these broad classes, the multiple interests and conflicting loyalties—family, professional, religious, regional, national, racial—that dissipate the common class interest. In short, his definition does violence at once to the simplicities and the complexities: the basic uniformities of all lives in a common environment, and the immense heterogeneity of individual or group interests and purposes.

This is to repeat the common observation that Marx had an archaic, metaphysician's psychology. Although he had a profound insight into the devious rationalizations and pretenses of material self-interest, he greatly overestimated the rationality of self-interest, greatly underestimated the powerful nonrational and irrational interests that make the economic man so seldom an enlightened or dependable political man. This deficiency is understandable in a philosopher, especially one who lived and wrote in exile. It is nevertheless grievous in a social philosopher, especially one who insists on uniting theory and practice. In this aspect, the doctrine of economic determinism is the inverse of the old religious fallacy, that it does no good to change institutions until you have first purified the hearts of men; Marx assumes that historic processes will automatically change institutions and that the change will as automatically fit men for the classless society. And so we come again to the wishful element of Marxism. Capitalism becomes the universal scapegoat, the cause of war, poverty, ignorance, crime, and all the other social evils that in the eighteenth century were attributed to the demon Superstition; Engels declares it "self-evident" that the end of capitalism will mean the end of all public and private prostitution. The Worker becomes the universal redeemer, with a corner on vigor and virtue; Marx tells us that the Worker's revolution will be unlike all other revolutions, and that his ideal destiny is his inevitable destination.

Now an incidentally curious limitation of Marxist analysis is that it seems always to stop short at the classless society—with the implication not so much that even the dialectical materialist cannot foresee the whole of history as that the dialectical processes of history come to an end. The established classless society pictured by Marx and Engels, and by many followers today, is pretty much an eternal, static perfection; they may allow for little disagreements over keeping all the wheels going around, but they suggest nothing like fundamental internal contradictions. In a way their Utopia is logical: when the class struggle that has dominated all history comes to an end, there will be no opposites to interpenetrate, no negations to be negated, no higher synthesis to reach. It is nevertheless rather incredible. And if one can hardly ask Marxists, as practical reformers, to begin studying the problems of the classless society, still their practical effectiveness as reformers has been limited by this undialectical attitude. In general, they have failed to allow for the unforeseen by-products, overemphases, and internal contradictions involved in logical as well as social structures. In particular, they failed to study the manifest internal contradictions in the U. S. S. R., and hence were later driven to the extremes of a bitter disillusionment with Stalin's Russia or a blind, fanatical defense of it.

Here is the most damaging criticism of Marxism, by its own standards. All its limitations as a sociology have led to its failures as a political program. They explain the serious errors in Marxist strategy: a gross overrating of the revolutionary zeal and force of the proletariat, a miscalculation of the strength and the trend of the middle class, a neglect of national and race feeling—a neglect of all the powerful irrational sentiments that Hitler has brilliantly exploited, and has also called out in opposition. The fault, generally, has been the fatal confidence in impersonal forces. Hitler's rise to power can be explained by natural forces; the exercise of his enormous power has been shrewdly calculated, determined by natural circumstances; but it has also been determined by a personal obsession, a mania, and the consequences of his individual will—or miscalculation—may now safely be called world shaking. Similarly the success or failure of the Soviet has hinged largely on the ability of the little group of men who made up the Gosplan, and then on Stalin's momentous shifts of policy—the "new formulas" that might be explained and justified, after the event, but were so far from being an "inevitable" event that experts in dialectical materialism were the least prepared for them.

The Marxist may answer that what shakes the world does not always move it, or that in a long enough view of history the equivalent of all this would have happened anyway. "Equivalent" is a dangerous word in human affairs, however, and a harsh word to the millions who now suffer from the shaking. And here, finally, we come to the reasons why the debacle of Marxism—at least outside of Russia—was, as Sidney Hook has called it, "a colossal *moral* failure." It was a failure in courage as well as intelligence; it was, above all, a failure in idealism. Because of their dogmas, Marxists corrupted the ideals that are required to justify their aims and to provide the force to realize these aims. They became in effect profound reactionaries.

The dogma of economic determinism finds no room for the small but very precious gains represented by civilization, the permanent residue left by every culture. Marxists have accordingly talked as if the ideas of tolerance and freedom were merely "petty bourgeois categories," and the aversion to brutal force were a symptom of liberal anemia; whereas such ideas are as old as civilization, and the very index of our precious heritage. Marx himself had an ideal faith, to repeat; in his thought there are seeds of a riper humanism. Yet the common fruits have been bigotry, fanaticism, downright inhumanity. Nor is this surprising. With its assumption of fixed ends, his program is a religious, not a scientific solution. It has therefore incorporated the Jesuitical principle that the end justifies the means, a principle that has been used to justify all the most barbarous, ferocious policies in history. On this ground everything goes—and the ruthless is likely to go best. Hence, more particularly, Marxism has meant a deliberate appeal to the method of violence, a glorification of revolution. Lenin noted proudly that Engels as well as Marx—and Engels was the more poised, humane philosopher—wrote a "veritable panegyric" on the bloody revolution to come. Even the more temperate Marxists, however—those who resign themselves to the bloody revolution as a regrettable necessity—cannot be excused for their supine surrender of the natural human effort to ameliorate and control. Meanwhile they did all in their power to *create* a revolutionary situation. They exploited the most delicate, honorable scruples of intelligent men of good will, the fear of being unrealistic and irresponsible. They have fixated a revolutionary psychology, the ruthless attitudes of the battle to the death, as the norm of all intelligent, solvent thinking. They have been merely contemptuous of the tolerance that is finally the necessary condition of thought, the necessary condition of civilization.

About Karl Marx himself, this is by no means the last word. These are logical consequences of certain of his ideas, but not the inevitable or the total consequence of his thought. Nevertheless the critic has to consider them, for they pervade the whole atmosphere in which literature has been written and judged. And for the critic there remains a last consideration, of the uses of dialectical materialism for his special purposes.

These uses are again too easily taken for granted. Marxists have denied so many evident truths that we may forget the evident truths they have discovered or reaffirmed. They have offered many valuable insights into capitalistic culture: how its Protestant religion, its individualistic philosophy, and its mechanistic science—all reflected in its literature—were conditioned by economic forces, and how the increasing restlessness and insurgency in all forms of its art followed the accelerated tempo of industrial revolution and the increasing loss of control of social relationships. They have offered a valuable criticism of the many other approaches—impressionistic, Freudian, semantic, Crocean—that still neglect the social context of all esthetic experience and value judgment. In general, their approach is indispensable to a full understanding of the conditions out of which literature inevitably grows and the ulterior purposes it inevitably serves.

No less apparent, however, are the limitations of Marxist criticism. Maxim Gorky once lamented that despite the excellent ideological equipment of revolutionary critics, "something seems to deter them from stating with the utmost clarity and simplicity the science of dialectical materialism as applied to questions of art." Actually there has been no dearth of simplicity; there is no need of reviewing the traffic in crude formulas, labels, and slogans that has passed for Marxist criticism. (The plight of Michael Gold is perhaps the classic example: finding the gaiety of Gilbert and Sullivan irresistible, even though Gilbert was a bourgeois Tory, he had to ease his conscience by sending in a hurry call for some proletarian operettas.) The more liberal interpreters accordingly felt obliged to make laborious demonstrations of the elementary. James Farrell wrote *A Note on Literary Criticism* because he was "forced to the conclusion" that the past is neither dead nor damned, and that its literature does after all have "some degree of esthetic and objective validity"; he clinched his case with quotations from Marx and Lenin, who as authorities in these matters have no more standing than the Pope. But even the more subtle, penetrating Marxists, and the many well-known critics who for a time were fellow travelers, have usually tried to explain too much by the economic interpretation, and have therefore suppressed too much, given a distorted view of literature both past and present.

So it may be worth repeating that this interpretation explains only certain *conditions* of cultural phenomena, not their cause or essence. It might explain a change in religious or esthetic practice; it cannot explain faith or the creative impulse itself. Art expresses much more than class interest, gives back much more than economic pressure, and its enduring values can never be understood in these terms. At best the Marxists, like the Freudians, throw light only on what great writers have in common with the mediocre. In *Illusion and Reality* Christopher Caudwell himself gave away the inadequacy of this approach when he declared that it is the essential task of esthetics—which as "mere appreciation" he had distinguished from literary criticism—"to rank Herrick below Milton, and Shakespeare above either, and explain in rich and complex detail why and how they differ." In a word, Marxist analysis is not the great pioneering principle that will open up a glorious new epoch in literary criticism. It is another perspective, another tool, another means to a much larger and different end; and it will be still more useful when it is itself seen more clearly in perspective, sharpened and supplemented by other tools, more thoroughly integrated with the whole literary tradition instead of substituted for this tradition.

Altogether, then, we have a social theory that remains a significant contribution even when it is only a crude approximation. It is perhaps better, in a way, that we have had so rigorous a development of the logic of a single principle, for we have not only a clearer view of the limitations of such a procedure, the dangers that social science must guard against, but a sharper perception of some important truths that must be incorporated in any valid social science. Nevertheless this principle can be applied more subtly and flexibly. The inconsistencies that make the text seem as elastic and ambiguous as Scripture are also the leads to a broader, suppler analysis. In such an analysis a great deal has to be sacrificed: the fixed ends, the ultimate cause and the inevitable effects, the guaranteed dividends for all policies in the mutual, the wonder-working power generally. The result will not look much like Marx's system. But much can be gained, and in the direction of the realism to which Marxists aspire: a grasp of still other important relationships, a clearer view of possibilities and contingencies, a broader basis for efforts at control. The hardy empiricist, like the scientist, can give up an absolute determinism without losing his determination, and in social affairs has a far better chance of foreseeing probabilities once he has surrendered the comforting illusion of certainties.

3. PARETO: THE MIND AND SOCIETY

"THE history of thought for the last eighty years," wrote Malcolm Cowley in his postscript to *The Books That Changed Our Minds,* "might be centered around the attack on the abstract conception of the Reasoning Man as the basis of social philosophy." By all odds the most sustained, systematic, and devastating attack has been made by Vilfredo Pareto. Nevertheless Cowley reports that he and his associates quickly agreed to exclude *The Mind and Society* from their symposium on the revolutionary books; and their decision is not surprising. As Pareto has become identified with Fascism, his attack is too revolutionary even for contemporary tastes. Cowley firmly believes "in democratic control and in the possibility of social progress"; though he joins the attack on the Reasoning Man, at bottom he wants to hang on to this concept, or at least to the faith behind it. So do I. Yet for this reason I believe that there is no better discipline for humanists than a close reading of Pareto's work. No thinker today more effectively compels us to reëxamine our premises or our gods, to return to the facts we are all so insistent upon facing. It is unfortunate that on some important matters he has *not* changed our minds.

Fundamentally, indeed, the thought of Pareto is very simple; when translated into everyday language, it is neither horrendous nor novel. As he himself constantly insisted, he was "doing nothing more than giving scientific form to ideas that are more or less vaguely present in the minds of all or almost all men, ideas that many writers have stated more or less clearly, and which facts without number do not permit us to ignore." His cardinal doctrine is that most human behavior is nonlogical (not necessarily *il*logical); in other words, he is investigating what men have always taken for granted as "only human nature." The basic uniformities of human nature, the elemental drives, he calls "residues." One of these is the itch to justify all conduct on logical grounds; so all the residues manifest themselves in theories or faiths that he calls "derivations." These rationalizations are the substance of our supposed knowledge.

> The history of social institutions has been a history of derivations, oftentimes the history of mere patter. The history of theologies has been offered as the history of religions; the history of ethical theories, as the history of morals; the history of political theories, as the history of political institutions.

Hence Pareto sought to analyze and classify the elemental social stuff, never leaving the field of observation for the happy hunting ground of ethics or metaphysics. By the rigorous application of the logico-experimental method, he hoped "to construct a system of sociology on the model of celestial mechanics."

Now, even celestial mechanics makes a dubious model for sociology. Although there is plenty of system in Pareto, the amount of science is debatable. Disciples have hailed his classification of the residues and derivations as his great original contribution, and it is indeed beautifully neat, with all the cozy categories and diagrams dear to the Latin mind. But it also seems arbitrary and it is not clearly useful for further inquiry. To my knowledge, the disciples have simply hailed the residues; they have made little constructive use of them or done little to put them in scientific fighting trim. For the layman, at any rate, Pareto's value lies chiefly in his destructive criticism: his remorseless analysis of other "scientific" sociologies, his unerring detection of the faintest whiff of sentiment where only logic is supposed to be, his ruthless exposure of all the dubious assumptions and pretensions of social thinking. He is accordingly strong medicine; taken straight, he is likely to corrode all idealism, make dead soldiers out of sick liberals. So it is important first to look into his own assumptions and see through his own pretensions.

Aside from his unconvincing method of proving that his system is scientific, which in practice is usually a scornful demonstration that all other sociologists are unscientific, one has reason to question Pareto's objectivity. This scientist who prides himself on investigating only what men do, not what they ought to do, is remarkably lavish with invective. Because ideal sentiments have been proclaimed as "natural laws" or dictates of "right reason," he is fiercely contemptuous of them even as ideals or sentiments. Whenever he mentions liberals, democrats, humanitarians, reformers, his voice gets shrill. He can even be a little ludicrous, as in his apparent obsession with people who would ban the sale of obscene postcards. On the whole, he displays an obvious bias that explains his reputation as a Fascist philosopher. His admiration is all for the Napoleons, the Bismarcks, the "strong men" unencumbered with scruples or tender sentiments; his contempt is all for the "spineless individuals" who cling to the ideals of liberty, peace, solidarity, progress. And in the excitement of whacking heads he grotesquely distorts even the facts he reveres. Anyone he despises, from Louis XVI to the politicians who governed England before the first World War, becomes *ipso facto* a *"fanatic humanitarian mystic."*

This bias is most damaging in the last volume of *The Mind and Society,* in which Pareto begins to deduce the practical consequences of the "experimental uniformities" he has discovered in three volumes of induction. Hitherto he had generally managed to confine his prejudices to his copious footnotes; the text of his analysis is impersonal enough. Now these prejudices run wild and produce a rank crop of inconsistencies, unproved assertions, and garbled facts. Invariably "experience shows" what he wants it to show. Throughout his discussion of the complex problem of social utility, he insists that in the absence of a common denominator for the heterogeneous utilities of individuals and groups, and above all of any scientific theory about proper objectives, no solution is as yet possible; nevertheless he is most prodigal with recommendations, most positive in praise and abuse of leaders. Because of his fierce resolve to eschew ethical considerations, his judgments are all conscientiously nonethical —as in his recommendation of the use of force and the policies of "lions" like Bismarck; nevertheless they carry him far beyond "experimental reality." Has Germany in the long run profited more by the accomplishments of Bismarck than it might have under a less ruthless ruler? Or if Germany has, has Europe, with whose welfare its own is ultimately tied up? I should not give a categorical answer. But neither should Pareto.*

These extravagances jolt the reader into an awareness of all that Pareto simplifies or suppresses in order to keep his diagrams neat. In his analysis of the residues determining the social equilibrium, for example, he entirely ignores his Class IV residues. These are the ones "connected with sociality"; they include "self-sacrifice for the good of others" and especially the "need of group approbation," a "very powerful sentiment" on which "human society may be said to rest"; but all these are potentially handsomer sentiments and lend themselves to idealisms. Again, because most thinkers greatly overrate the influence of derivations, Pareto must insist not only that behavior governs theory more than theory governs behavior, but that the power of theory is negligible; he overlooks the strong hold of specific dogmas or faiths, because of which similar sentiments give rise to very different behavior in different societies. He overlooks as well the force of definitely logical behavior, and above all the profound influence of science today upon the ways of life, the habits of thought, and ultimately the sentiments of men. "Reason," he admits, "is coming to play a more and more important role in

* Today the successes of Hitler may force one to seek refuge "in the long run"; but there is also the spectacle of the unhappy Mussolini, who has declared himself a pupil of Pareto. His *Autobiography* now makes good rereading, especially the passages in which he speaks complacently of all he has done toward "broadening the world's concept of its duty to youth." Italy's youth he had trained through a discipline "rigid but gay," and so he had developed a military spirit that was "lively" but not "aggressive." "They are made to see a sure vision of the future." In fairness to Pareto, however, it should be added that this lion already betrayed signs of softness: "A sentimental motif [has] stamped itself upon my soul and upon the grateful spirit of the nation."

human activity"; but why, in what ways, to what effect, he does not say. Meanwhile all his sweeping generalizations are based upon analyses of only one kind of behavior.

Even within its scope, however, there is a serious weakness in Pareto's inductive analysis—in the foundation itself of his system of sociology. F. S. C. Northrop has remarked that Pareto's procedure was a rather strange one for a scientist; all his "facts" he got from classical texts and newspaper clippings, never leaving his armchair for firsthand observation. Apparently he seldom bothered even to consult the work of sociologists, psychologists, and anthropologists who had made such observation, for his few references to them are chiefly sneering. (Curiously, he never mentions Freud, although much of his analysis of nonlogical behavior is clearly in line with Freud's theories of the unconscious.) But at best his inferences are still questionable. Pareto provides no clear criterion by which to determine uniformities, no logico-experimental test by which to verify his residues. One has to take them on faith, precisely as one takes the different lists of instincts drawn up by McDougall and used by Graham Wallas in *The Great Society*. Ultimately his whole "scientific system" is inspired guesswork.

Ultimately, in fact, it is not even clear what a "residue" is. It is a "manifestation" of something or other, but on the processes of manifestation Pareto is silent. Evidently it corresponds to biological instincts, innate constants, else he could not generalize and predict so freely. Accordingly he tells us again and again that residues change very little and very slowly, if at all. But now embarrassing questions arise. For one thing, the many parallels that Pareto cites as evidence of innate uniformities are by anthropologists usually explained as the result of imitation or historical diffusion. Again, he fails to explain why these biological constants assume such diverse social forms, which in turn so greatly influence behavior. He merely mentions the "derivatives"—the modes of behavior as apart from the theories or derivations that "manifest" residues; this central province of the mores he left to Sumner, who explored it with a cool objectivity that makes Pareto seem an evangelist—and with whose work he was, characteristically, unacquainted. Furthermore, there remains the puzzling fact that these basic uniformities do vary, from class to class, country to country, age to age. Sometimes, indeed, they appear to change very rapidly. Pareto refers the miserable showing that Prussia made before Napoleon in 1800 to an intrinsic weakness in his Class II residues ("group-persistences"); but a little later he says that in 1870 Prussia was infinitely superior to France because it was so strong in these residues. In general, Pareto shows the consequences of such variations in his ingenious analyses of contrasting societies, he points to the widespread decay of certain primitive sentiments and the growth of humane sentiments in the modern world, he declares that the whole class of "group-persistences" is losing ground to the "instinct for combinations," which is the progressive factor in society; but precisely how or why these changes occur, and how much then is actually uniform, he does not say. He simply returns to his theme song, that residues are "very very hard to modify" but derivations stretch like rubber bands; and then he runs to his footnotes to fume at thinkers whose hopes for a better society ultimately rest on just this fact that residues do change, with its implication that "human nature" may be further educated.

Here is again the tragic paradox of contemporary thought. In their new awareness of the unconscious forces that mold society, thinkers forget that this awareness is itself intelligent and can inform conscious purpose; and so we have Pareto, like Spengler, brilliantly exercising his intellect to disparage the power of intellect, brilliantly exploiting the accumulated resources of civilization to prove that progress is illusory. In effect he cuts off the head of the social organism.

All in all, upon completing Pareto's prodigious work we may feel like rubbing our eyes, as if we had just come down from Mars, when we return to the familiar, untidy, hazardous world in which we perforce live out our lives—the world where morals and ideals are not ludicrous examples of slovenly thinking but vital concerns. Values are indeed implicit in all his judgments, and become explicit when he approaches the problem of social utility. But utility involves such intangibles as happiness, spiritual as well as material welfare, the full life—utilities at once real and ideal, which as yet cannot be measured or verified but cannot for that reason be ignored; and he admits them chiefly as opportunities for ridicule. In practice he restricts himself to the measure of "prosperity," because with this he can get some statistical indices, enjoy the illusion of an exactness comparable to that of celestial mechanics.

Yet Pareto was too honest really to enjoy this illusion. Eagerly he introduced mathematical formulas whenever possible, just to show how exciting it would be to get quantitative measurements; but he always added, mournfully, that we cannot use these formulas: the unknowns and immeasurables are too many, the interdependencies are too complex, the objectives are still beyond the scope of science. After all his rages and his contradictions, he soberly returned to the stubborn facts. And here we approach the positive value of his thought. Even after it has been pruned of its shaggier extravagances, supplemented as he hoped it would be, I doubt that it will be the cornerstone of the sociology of the future—at least until we have learned much more about the nature and force of all the mysterious "psychic states" that Pareto imaginatively reconstructed from his armchair reading. Nevertheless he contributed a great deal by surveying and preparing the ground, defining the essential problems. He had a clearer, steadier, more comprehensive view than any sociologist before him of the intricate interrelationships, the multiple causes and effects, the endless actions and counteractions, and all the irrational factors and unknown quantities that constitute the subject matter of this science. He is perhaps the best illustration of what Mannheim calls "interdependent thinking."

It is impossible to do justice here to the many penetrating observations strewn throughout the two thousand dense pages of *The Mind and Society,* the specific analyses of the form and substance of legal, religious, political, and other social institutions, or even to all of Pareto's valuable concepts. His theory of class-circulation, for instance, may explain less about social change than does the theory of class-struggle, but class circulation is especially important in democratic societies, conditioning the class struggle to an extent unrecognized by Marx; and Pareto is sounder in that he always regards it as but *one* factor in a complex, is *content* to explain less by it. For the layman, however, his chief value stems from his premise that nonlogical behavior is not an incidental aberration but the fundamental social fact, and from his searching analysis (not his elaborate classification) of the deep sentiments and all the pseudological forms they assume.

Thus Pareto repeatedly illustrates the universal confusion of the sign or index of a thing with the thing itself. A sentiment manifests itself in forms *a, b, c,* and *d;* men are prone to regard *d* as the *cause* of *a, b,* and *c,* and even of the sentiment behind them; and then they try to destroy the whole context by attacking *d.* This kind of futility is conspicuous in the efforts to make a whole citizenry moral by laws imposing prohibition or sex hypocrisy, but it is at the heart of almost all the theories about "what this country needs." Efforts at reform so often end in preposterous simplicities because they are efforts merely to *re-form,* but are nevertheless based on the faith that changing the form will automatically change the substance of society. At the same time, these efforts are themselves expressions of sentiments; and Pareto constantly exposes all the social thinkers who take their own sentiments for truths and dismiss those of which they disapprove as outworn prejudices, who at that seldom state or even clearly conceive the norms underlying their judgments of right and wrong, and whose final argument is an appeal to what "every honest man must admit" or "every reasonable man will see." He shows, too, how they habitually simplify the problem of social utility, assuming that a certain principle or policy is absolutely and under all conditions necessarily beneficial. And in all this his very bias is an aid, for it leads him to dissect with most thoroughness the sacred cows of contemporary thought—the "modern sense" of this or that, the "spirit of broad humanity," and all the slogans and passwords that touch off fashionable sentiments and substitute good will for clear thought. A study of his work can be an excellent setting-up exercise for working off layers of fatty sentiment, giving tone and toughness to the muscles of the mind.

More specifically, Pareto's approach led him to a clear realization of the purely linguistic confusion along the whole intellectual front. He not only was a pioneer of semantics but remains one of its most acute, effective practitioners; his entire analysis of the derivations is an analysis of the emotive use and logical abuse of language. Hence his pains to invent a complete, generally unlovely terminology of his own. If he was at times entranced by the manipulation of this new verbal machinery, his principle was nevertheless sound: only by the strict use of a technical vocabulary can sociology hope to become a science instead of a fancy way of ringing changes on old sentiments. Yet also implicit in his thought is the soundest criticism of the excesses of the semantic reformers, as of rationalists and positivists generally; for he also perceived the practical necessity and the value of the habits of thought and speech that they simply condemn. He not only delighted in showing that the most logical derivations may be the unsoundest, because the most consistent, thorough developments of unsound premises; he asserted that irrational or even absurd beliefs may nevertheless be socially useful.

Among Pareto's favorite objects for exercises in derision, in fact, is the "goddess Science"; among his favorite challenges is the flat statement that "the worship of Reason may stand on a par with any other religious cult, fetishism not excepted." This explains the fate he foresaw and took a characteristically ironic pleasure in: he whose life work was a monument to pure science would be especially obnoxious to its most devout disciples. (This also explains some of his disagreeable teachings, for he seems to have leaned over backwards in his efforts to be objective and realistic; his friends have reported that this advocate of the use of force was at heart a "profound pacifist," and during the World War was indignant at the pacifists who turned patriots.) Fundamentally, however, his position here is sound enough. He maintains that man's behavior is governed primarily by sentiments, that these sentiments are not necessarily contemptible, and that it is folly to ignore or despise them in theorizing about human society or trying to promote its welfare; he also insists upon seeing these sentiments for what they are and not permitting them to masquerade as philosophical or scientific thought. What he attacks, accordingly, is not reason or science but their capitalized abstractions, with their retinue of false appearances, fantastic claims, and tyrannical demands. In short, Pareto belongs with those thinkers who gratefully accept the scientific rationale but also recognize the need of supplementing it, perceiving that it has slighted, and alone cannot satisfy, the deep nonlogical purposes and pieties that are the necessary cornerstone of any social edifice.

For this reason Pareto's thought has especial interest for the literary critic. In general, his criticism of social historians holds as well for the historians of literature; they are no less prone to mistake the history of esthetic theory for the history of artistic performance, and are given to even fancier patter as they run the gamut of derivations. He also warns incidentally against the tendency to generalize freely about a whole society on the basis of the traits and attitudes of its elite. (The distress of sensitive contemporaries amid the blaze and blare of popular taste has been aggravated as they read in their histories of the glamorous Elizabethan age, presumably typified by Shakespeare, Raleigh, and Sidney, or of the ultracivilized eighteenth century, in which ladies and gentlemen perpetually held high discourse and paraded their fine manners, wit, and charm—forgetting that the citizenry of all ages have had an inveterate preference for the mediocre, and that by its nature the mediocre is perishable and leaves few records.) But more important is the whole approach to nonlogical behavior. "Literature satisfies the human hunger for combining residues," Pareto writes, "and that need is left unsatisfied by science." Broadly speaking, literature finds its primary subject matter in sentiments, its ultimate purpose in an appeal to sentiments. Here is another perspective on the problem of "beliefs" in literature, and of how great literature outlives the specific ideas it embodies. These are derivations, and whether or not they are merely "pseudo-logic" (or "pseudo-statements," as I. A. Richards calls them) they do not form the enduring value and significance of literature. This comes instead from the command of the relatively ageless residues behind them; sentiment gives the richest meaning to concrete perceptions. Hence a writer may satisfy his own age by a vivid expression of its cherished derivations—its particular causes and faiths—but he will not satisfy a later age, or profoundly affect even his own, unless he has penetrated to the deep springs of behavior, expressed much more than these changing forms. And Pareto accordingly recognized that the linguistic habits fatal to logical analysis are essential to creative literature. It is the writer's business to identify words with things, emotion with thought.

Implicit in Pareto's system, finally, is a complete statement of the limitations of Marxist theory and the excesses of Marxist practice. Pareto explicitly criticizes the doctrine of economic determinism. From the complex of forces determining the social equilibrium—such as race, the state of knowledge, climate and natural resources, contact or clash with other societies—he singles out, as most convenient and important for analysis, four internal forces: interests, residues, derivations, and class circulation. He acknowledges that Marx's analysis of the effect of economic interests upon the other elements is especially valuable for the modern world, given an industrial society in which these interests are preponderant. Nevertheless he makes clear the fatal error of Marx in isolating this one factor as primary, disregarding the constant, simultaneous counteraction and interaction. His own system is not only more objective; it is more comprehensive, more consistently organic, more genuinely sociological.

Yet Marx's is the more dynamic. Another fundamental difference between the two thinkers is marked by the evolutionary concept. Roughly, it is the age-old difference between the "ideological" mode and the "utopian" mode, the philosophy of being and the philosophy of becoming. Marx, stemming from Hegel and an age thrilled by the possibilities implied in evolutionary theory, naturally adopted a philosophy of becoming. His system is in this respect more scientific than Pareto's, at least by current theory, and in the long view of human history has a more solid basis in fact. Pareto, seeking to classify uniformities and also reacting, like Spengler, against the simple gospels of progress inspired by evolutionary theory, as naturally adopted a philosophy of being. Hence he drew up his historical parallels to prove that change is not forward but cyclical, and by a specific parallel between the modern world and the declining Roman empire arrived at another of the gloomy forecasts that so fascinate harried contemporaries. Nevertheless Marx had no more scientific warrant for predicting the end of the evolutionary process than Pareto for predicting the end of the present cycle. Each worked his theory for more than it was worth, each ignored much of the plain truth in the other; the partisans of either can easily demolish the other. Ultimately, both social philosophies are metaphorical.

The argument for the Marxist philosophy as the wiser choice (though by no means the only possible alternative) must accordingly rest on pragmatic grounds. One may say that it has greater social utility in stimulating energy, providing goals for an incorrigibly purposeful and ethical animal, affording a richer diet for human interests and sentiments; in Pareto's own language, it is more satisfying to the progressive "instinct for combinations" that he declares is gaining ground today. In general, I should say that there is a broader basis of objective truth in Pareto, a firmer basis of functional value in Marx. Thus Pareto is the better guide to the whole period between the World Wars, and especially to the recent history of the Soviet; if liberals had taken to heart his analysis of the forces of sentiment, the ways of leaders and masses, they would have been prepared for the developments that have so profoundly shocked and disillusioned them. But Pareto does not explain or even allow for such a mighty experiment as the Soviet; Marx made possible this experiment, providing the essentials of both motive and method. Similarly, for the literary critic, Pareto more clearly defines the fundamental subject matter of literature, the source of its enduring values; Marx has stimulated writers by suggesting a more specific subject and purpose, an immediate source of values. For all men, social life remains an experiment in values. Marx offers the more fruitful hypothesis for a possible end in this experiment; Pareto throws a steadier light on its given terms.

4. ANTHROPOLOGY

In exploring the wilderness of sociological theory, one is relieved to come upon the clearing of anthropology, where he may contemplate a modest but carefully tilled and weeded body of thought. Anthropologists have to contend, indeed, with the same underlying difficulties as the sociologist. They too cannot employ the ordinary scientific devices of measurement and cause and effect; they cannot isolate and experimentally control any single factor. But they also have some obvious advantages. Although sociologists may study limited cultural areas—regional, class, professional, etc.—as the Lynds did in *Middletown,* all these areas overlap, all tracks crisscross and tangle in the immense, intricate, heterogeneous sprawl of modern civilization; in the relative simplicity of primitive societies, anthropologists can more easily make out both the interrelationships and the society as a whole. Furthermore, they can work in something like a laboratory, concentrate on scientific fieldwork instead of historical guesswork. They are not called upon to act as missionaries and reconstruct their materials. Hence they have usually conducted their inquiries in a more objective, dispassionate spirit and are generally accepted, even by purists, as genuine scientists. They have been able to demonstrate most conclusively the peculiarity and the importance of the subject matter of social science, showing how man is above all a "culture-building animal," and how a given culture molds his entire consciousness and behavior.

More immediately, anthropology affords a striking example of the development, operation, and practical value of the organic concept. Under the spell of simplified evolutionary theory, Morgan, Tylor, and other early investigators pictured a more or less steady unilinear development from a generic primitive to a civilized state, with more or less definite levels in cultural institutions; as economy progressed from the stone to the bronze to the iron age, so did religion progress from animism to polytheism to monotheism, with moral ideas following dutifully in line. It was a comfortable view, enabling scholars to root for their beloved origins, missionaries to convert the heathen, reformers to tackle optimistically the residues of primitive customs, and all men to feel complacent about their superior institutions. Similarly anthropologists made collections of museum specimens of the primitive, like *The Golden Bough,* interesting and valuable but monstrous miscellanies, no more a science than is a museum. Even Sumner assembled all his exhibits of the mores in separate showcases, where they could be conveniently catalogued but could not be studied in their relations. Today, however, anthropologists stress not only the amazing diversity of the cultures that have been lumped together as "primitive," but the necessity of studying any one culture as a totality. They aim to explain all anthropological facts, says Malinowski, "by their function, by the part which they play within the integral system of culture, by the manner in which they are related to each other within the system, and by the manner in which this system is related to the physical surroundings."

Perhaps the best example of this concept in operation is Ruth Benedict's *Patterns of Culture.* Miss Benedict assumes that the significant unit is not the institution but the whole pattern, the "cultural configuration." This is not always easy to make out. Some primitive societies are loosely integrated, and even the neatest patterns have bulges or gaps; all are complicated by cultural diffusion, the influence of surrounding cultures in space and time. A perfect unity exists only in the purely ideal state of a Hegel or the ideally material state of a Marx. Nevertheless an integrative force is always at work, the cultural whole is still more than the sum of its parts. To illustrate, Miss Benedict analyzes three cultures as remarkably well integrated as remarkably diverse: the Zuñi in New Mexico, the Dobuan in the South Seas, and the Kwakiutl on the Northwest coast of America.

Let us take a look at the Dobuans. Their society is organized by a permanent, universal animosity and malevolence; sullen suspicion and resentment are their chief motives, ill will and treachery their chief virtues, and the whole culture "provides extravagant techniques and elaborate occasions for such behavior." Thus marriage in Dobu is formally initiated by a hostile act of the mother-in-law, who publicly traps the youth who has been sleeping with her daughter; this ceremony of betrothal leads to further humiliating ceremonies, until finally the sullen victim is introduced into a clan in which he will always be considered an alien, and linked to as sullen a partner who will be as pleased to commit adultery whenever possible. Success in agriculture, the basis of Dobuan economy, is a tacit confession of theft, supposedly having nothing to do with planting or natural care; it means that by a recital of charms and countercharms one has attracted his neighbor's yams into his own field while preventing his own yams from wandering into other fields. Disease is also attributed to the malevolence of some fellow citizen: most persons own from one to five charms that are specifics for particular diseases, and some fortunate souls have a monopoly of a certain disease. Similarly there are no forms of legality or political organization, for lawlessness is highly ritualized and anything one can get away with is respected. In short, all institutions are shaped and all existence is dominated by this central obsession with malignity. No provision is made for kindness, friendliness, laughter, joy; any show of politeness or consideration is the surest sign of deep treachery. Miss Benedict concludes:

> The Dobuan lives out without repression man's worst nightmares of the ill will of the universe, and according to his view of life virtue consists in selecting a victim upon whom he can vent the malignancy he attributes alike to human society and to the powers of nature. All existence appears to him as a cutthroat struggle in which deadly antagonists are pitted against one another in a contest for each one of the goods of life. Suspicion and cruelty are his trusted weapons in the strife and he gives no mercy, as he asks none.

Almost as unlovely, though in a different way, is the Kwakiutl culture, whose central motive is a will to superiority that amounts to megalomania and that expresses itself in fantastic arrogance and destructive waste; the whole economic and social organization is bent to the purpose of asserting one's own greatness and humiliating all others, with institutionalized life being chiefly an exchange of affronts, and victory-shame the only recognized gamut of emotion. (Even the accepted form of homage to their gods is not to reverence but to shame or even kill them, in order to secure supernatural powers—which would seem to be the last word in the practice of returning to God the compliment of creation in one's own image.) In striking contrast with both of these, however, is the Zuñi culture, a remarkably ceremonialized culture dominated by the ideas of moderation and coöperation: it plays down all sensory and emotional excitement, all personal reference, all individual acquisitiveness and authority. These contrasts are indeed unusually striking—and Miss Benedict has been accused of some exaggeration in detail. But her main thesis appears to be firmly established. Sexual, religious, economic, domestic, political, and other interests or activities are not self-contained areas of behavior, each with its own drives or laws. They are, rather, "occasions which any society may seize upon to express its important cultural intentions," and are again and again dominated by the whole configuration. All three of these cultures manage characteristically, for instance, the great occasion of death—but none treats it in what seems to us the "natural" way, as a grief situation. The Zuñi recognize it as an emergency or loss, but as far as possible minimize grief and terror; the Kwakiutl use it as another expression of their "cultural paranoia," regarding it as the paramount affront or shame and seeking some bizarre compensation (a common form is the gratuitous killing of someone equal in rank to the deceased); Dobu interprets it as treachery, an occasion for mutual recrimination, and punishes the blood kin of the surviving spouse. Likewise even physiological facts are not only socially interpreted but sometimes effectively suppressed. The "natural" turbulence of adolescence may be unrecognized in a culture and become actually unimportant to the child; Margaret Mead has shown that in Samoa a girl's puberty falls in a particularly peaceful, uninteresting stage of her life as a member of society.

Miss Benedict makes no effort to explain how or why these dominant attitudes develop or often fail to develop, how or why the patterns change; she offers little further generalization *about* the patterns. Following her very cautious mentor, Franz Boas, she shies away from comprehensive "laws." The caution is no doubt well advised. Nevertheless anthropologists are gathering further suggestions from comparative studies, getting at some principles of patterning.

Thus Margaret Mead has edited a group of studies of cooperation and competition among various primitive peoples. She finds that the emphasis upon one or the other depends upon the total emphasis of the culture, not simply upon material conditions or biological needs. But she also finds certain uniformities beneath the striking diversities. All the competitive societies are marked by a dependence upon individual initiative, a high valuation of property, a single scale of success, and a belief in arbitrary supernatural powers. All the coöperative societies are marked by far less emphasis upon individual initiative and power, a high degree of security for the individual, and a faith in an ordered universe; "social life goes on in response to a structural form expressed in a calendar, a court ritual, or a seasonal rhythm." In short, the anthropologist can find some correlations *between* cultures as well as within them.*

A different approach to significant generalization is marked out in *The Individual and His Society,* in which Abram Kardiner psychoanalyzes the Tanala and Marquesan cultures on the basis of field reports by Ralph Linton. He turns to individual personality as the logical place to study cultural integration; we can know institutions only by understanding the behavior of the individual, for he is simultaneously their creator, their carrier, and their creature. Kardiner formulates the concept of a "basic personality structure." All men are born helpless and for a long time are dependent; all have sexual appetites and other bodily needs; all wish to be effective, seek to master or control objects for utility or pleasure; all tend to hatred or rage when important needs are frustrated. Correspondingly, all cultures are fundamentally alike. They all have some form of family organization, some definite techniques of food getting, some basic disciplines, some standards for controlling mutual aggression, some distinctive life goals, etc. The basic personality structure, the frame within which the individual develops his character, is determined by the way a given culture meets, controls, or frustrates the universal needs. By analyzing this basic personality, the

* We have here another perspective on the confusions and conflicts of modern civilization. These owe not merely to cultural lag but to the different emphases of a society that has been at once fiercely competitive and unprecedentedly coöperative. Yet there is no necessary, absolute incongruity here. Among these primitive peoples competition does not always mean conflict, nor coöperation solidarity, and competition always presupposes a considerable degree of coöperation. At any rate, the more fundamental factor in our society is coöperation; there can be no industrialism without it. Industrialism imposes form, calendar, rhythm, and, like science, implies a faith in an ordered universe.

anthropologist can get a better idea of the fulfillments a culture permits, the repressions it demands, the anxieties it creates. Here Freud's concept of repression is extremely useful, since all cultures repress some impulses, and the neuroses that result when basic needs are persistently interfered with are the handiest key to the institutions creating the pressure. By a comparative study of the effects of different disciplines, the anthropologist may perform something like an experiment. Kardiner believes that a study of twenty or thirty cultures in this manner would enable at least an approximation of the laws governing the "psychodynamics of social change."

Meanwhile anthropology can at least serve as a court of appeals from all the too hasty legislation that social scientists draw up after studies of a single culture—their own. Classical economists are the notorious example.

> Political Economy or Economics [Marshall's sound definition reads] is a study of mankind in the ordinary business of life; it examines that part of individual and social action which is most closely connected with the attainment and with the use of material requisites of well-being.

In practice, however, Marshall did not stick to this definition: he studied capitalism, not mankind, and reduced economics to the phenomena of price. This is a legitimate practice, of course. But it does not authorize economists—and businessmen and politicians after them—to offer their "laws" as laws of human nature.

Here is again the fundamental limitation of the system of Marx: he presented an economic analysis of capitalism as a philosophy of all history. Anthropologists completely refute his principle of economic determinism. From the beginning, George Grant MacCurdy points out, the change from the stone age to the age of metals—the most revolutionary ever made by man—for a long time involved no radical change in his mode of living; innumerable field studies show extremely diverse cultures, on the American prairies as in the South Seas, growing out of the same material conditions of life; and all anthropology is a demonstration of the multifarious endings of common beginnings. The principle of class struggle explains still less in primitive cultures, in which class or caste takes innumerable forms but seldom one as simple and handy for the dialectician as the bourgeois-proletarian setup. Altogether, primitive life seems to have gone on with little regard for the logic of dialectical materialism.

Similarly anthropology makes clear the need of a historical, sociological psychology. Psychologists have commonly disregarded cultural influences and generalized about all human behavior on the basis of our particular kind of behavior. Freud is the conspicuous example. With his theory of instincts he adopted the theory that the child recapitulates primitive experience, another simplified view of evolution and mistaken view of the "primitive"; when he wanted an origin he invented a scientific myth, such as the prehistoric event that inaugurated the Œdipus complex. By inventions that cannot be verified, he persistently obscured the social influences that can be studied. Actually, Kardiner could find no signs of an Œdipus complex in Marquesa, where there is no taboo on the child's sexual desires; he found plenty in Tanala, where the child has to learn to behave himself more as he does in our society. This complex therefore appears to be the product of social institutions, not of instinct or racial memory. In general, anthropologists point to the necessity of studying all "abnormal" behavior in its social context, and of questioning the fixed criteria of the reality principle and the pleasure principle, "good" and "bad" ways of dealing with a situation. The Zuñi seem to be fostering neurotic behavior, for they run away from grief and all unpleasantness, make a habit and a virtue of the "escapist" principle. Nevertheless this principle seems to work with them; it is supported by all their institutions and attitudes, and the group as a whole gives a strong impression of mental health. In Dobu, on the other hand, the thoroughly maladjusted, abnormal person was a naturally friendly, kindly, honest fellow who did not try to give all his neighbors diseases or charm their yams away, but was glad to help anybody. In fact, the most successful and honored type in a given culture is likely to seem patently abnormal to another culture. As we regard the cataleptic priest of the savages or the Puritan divine, so may posterity regard our eminent money-grubbers and aggressive captains of industry.

Especially pertinent today is the evidence of anthropology concerning the racial theories popularized by Houston Stewart Chamberlain and Madison Grant, and now made the basis of Hitler's mythology. There is no pure Nordic or Aryan race anyway, and it would take some seventeen generations of strict inbreeding to get a strain that would breed true; but anthropologists can find no clear evidence for the superiority of any race, or for the advantage of purity of blood. Every civilized group we know of has been a hybrid.

The growth and spread of civilization [writes Ralph Linton] has gone on with a serene indifference to racial lines. All groups who have had an opportunity to acquire civilization have not only acquired it but also added to its content. Conversely, no group has been able to develop a rich or complex culture when it was isolated from outside contacts.

Likewise anthropologists discredit even the more harmless forms of the race hypothesis that were once popular among literary men. (Arnold, Carlyle, Meredith, Stendhal, Taine, Sainte-Beuve, and Renan, to name a few.) Arnold gave the case away when he appealed to Englishmen to give more prominence to their Celtic blood. If we can dictate to our blood, decide what strains merit special attention, then culture and not race is the ruling force; and this is the verdict of anthropology.

For social purposes generally, the chief value of anthropology is the light it throws on this all-important problem of what is culturally and what biologically determined, or in other words our old problem of the nature of human nature. Precisely what is given at birth we cannot say. But we can say that none of man's culture is carried in his genes, and that very little of his social behavior appears to be engrained in instinct. Other animals begin life better equipped and have little to learn; man has to be taught how to behave, and during the long period of dependency after birth learns fantastically diverse modes of behavior. There is scarcely a "law" of human nature that has not been ignored or repealed in some culture. Such basic needs as are given can be channelized, modified, elaborated, or combined in so many different ways that these permanent conditions are also virtually unlimited possibilities.

It is an extraordinarily simple matter [writes Miss Benedict] for tradition to take any occasion that the environment or the life cycle provides and use it to channel purposes generically unrelated. The particular character of the event may figure so slightly that the death of a child from mumps involves the killing of a completely unimplicated person. Or a girl's first menstruation involves the redistribution of practically all the property of a tribe.

The same situation may accordingly have now religious, now sexual, now economic associations—none of them more intrinsic or necessary than our habitual association of marriage with religious sacrament. In any event, a cultural institution is almost certain to go far beyond its original motive or drive as it becomes interrelated with other institutions and is shaped by the whole configuration. Least of all can the complex institutions of modern society be explained by simple biological causes. The profit motive of capitalism may be an outgrowth of the now deeply rooted habit of appropriating things, but it is not a necessary manifestation of an "acquisitive instinct."

In short, the instincts of common discourse are less the cause than the result of social institutions: a man is acquisitive *because* he can own property and acquire prestige with it. As Dewey holds, man is not primarily a creature of instinct or of reason; he is a creature of habit. Contrary to the belief of conservatives, instincts are easier to educate and modify than are institutionalized habits. Contrary to the hope of radicals, such habits are extremely difficult to eradicate. Children never resist education or innovation as do civilized adults; they may be full of disgusting impulses but they have few settled habits. The still very general tendency to assume inviolable laws of human nature, and to find these laws written in local customs, is itself evidence that the Idol of Idols—of the cave and of the tribe, in the theater and in the marketplace—is Habit.

The immediate moral here, then, is the necessity of taking really adequate account of cultural relativity. Although at bottom we all know that our ways are not universal or inevitable, we fall insensibly into the vulgar habit of assuming that they are necessary, right, and generally superior. We look upon such institutions as war as unfortunate relics of our savage past; whereas organized warfare is actually a late development in social evolution, and some savages have been simply incredulous when white men tried to describe this institution to them. We can see that early societies seldom dealt "realistically" with frustration, providing subjective satisfaction but only the illusion of objective control; we do not see that outside the scientific laboratory or the psychiatrist's office we are usually as unrealistic, cultivating the same illusion by prayer or punishment. (The Eskimo's idea of restitution, by which a murderer becomes husband of his victim's widow so that the family will not be deprived of a food getter, offhand seems more realistic than our custom of killing him in turn.) In general, all our institutionalized ways of treating the great life situations—the ageless cycle of birth, growth, reproduction, toil, and death—are not necessarily more efficient or more reasonable than the ways of primitive cultures.*

In such ways we may also see around even our basic modes of thought. Primitive thought was long considered merely rudimentary or childish, as if its animism proved a simple incapacity for real thinking. It reflects, indeed, no comprehension of natural laws as we conceive them. "Men mistook the order of their ideas for the order of nature," Frazer explained, "and hence imagined that the control which they have, or seem to have, over their thought, permitted them to have a corresponding control over things." But this is to say that they were aboriginal Hegelians. Lévy-Bruhl pointed out, moreover, that primitives often display considerable subtlety and ingenuity in working out their animistic scheme, with all its mystic identifications and exclusions, and that this scheme is at its best complete, coherent, consistent, and therefore "logical"—more so than the hodgepodge of science, superstition, and hearsay that is the ordinary man's scheme of thought today. Their indifference to abstract concepts is the very measure of their logical consistency. At any rate, this indifference emphasizes the "sociology of knowledge." It suggests the advisability of reviewing our heritage in the spirit of an anthropologist, looking upon the philosophy of Greece, say,

* The Kwakiutl culture is particularly suggestive for contemporaries, as an innocent burlesque of our system of private property and the profit motive. Their society was based on great possessions (including even areas of the Pacific Ocean), and their social existence was a tireless, ruthless competition for exclusive ownership—ownership of names, myths, titles, supernatural powers, and all kinds of prerogatives with little or no economic function. They manipulated their vast immaterial wealth with the aid of an elaborate monetary system, whose counters had as little intrinsic value as our paper money; and they could accumulate this wealth even more rapidly than our financiers, as they had the benefit of 100 per cent yearly interest rates. Similarly the gratifications they got from wealth were a fantastic exaggeration of the "conspicuous consumption" and "conspicuous waste" that Veblen described in our society. They humiliated a rival either by presenting him with more property than he could return with interest (as he was obliged to) or by destroying more property than he owned (since the destruction of an equal amount was the only way of avoiding shame). In short, by extending the basic motives and values of capitalism into all spheres of social life, making ideal virtues of business practices that we still condemn in private relations, the Kwakiutl nakedly expose the folly and the inhumanity that we too often accept as inevitable, or even right and good.

as the expression of a particular culture instead of the necessary cornerstone of thought. It supports the view of logic as a progressive discipline, dependent upon the state of inquiry as well as the structure of the human mind. It confirms the need of caution in fixing a priori, necessary categories of thought.

At the same time, one may question Lévy-Bruhl's assertion that primitives lack the category of cause and effect. He tells of the African natives who suddenly displayed an uncharacteristic enthusiasm for learning to read and write. They were being paid with credit slips that were cashed in at the company store, and they decided—with no intention of fraud or theft—that if they could learn to make marks on paper, they could get all the merchandise they wanted without having to do all the work. Here, Lévy-Bruhl concludes, was a simple failure to reason through cause and effect. Actually, it seems clear that they were reasoning so. They saw the marks as the "cause"; as with all their fetishes, they merely hit on the wrong cause. And such reservations are important, as safeguards against the easy assumption of utter relativity. For all its bewildering diversity, the spectacle of human culture calls for a redefinition, not a denial of our common humanity. There are indeed permanencies, as noted by Kardiner; there also has been an evolution of culture. To say that men cling to habits, stubbornly persist in nonrational or downright irrational behavior, is to state something like a universal law of human nature. To say that men nevertheless are able to criticize and change their habits, in the interest of more rational behavior, is to state the permanent possibility of a better society.

Hence we need to qualify and supplement Miss Benedict's analysis of the patterns of culture. Although she does not deny, she tends to ignore the natural and social continuities, and to present her cultures as incommensurable, the varieties of custom as equally valid. Although she does pass occasional judgments, describing customs as "desirable" or "undesirable," she does not define the norm implied in these judgments. She leaves too far behind the natural sciences that can help to locate the continuities and define the norm. Biology, she declares, provides the possibilities or tendencies; culture provides the situations that channel them. Nevertheless biology also shapes the situations and the channels, always acts on the patterns and in the long run selects from them. A culture can never be fully understood without a knowledge of the basic structure and functions of man, and without consideration of its natural environment.

These qualifications, however, only strengthen Miss Benedict's concluding argument, that the recognition of cultural relativity carries with it its own values. Sophisticates have usually made it an excuse for cynicism and nihilism, or a doctrine of despair. But it can also serve as grounds for hope, the means to a more realistic faith. Our cherished values not only should be able to bear examination but should gain by a comprehensive survey and criticism of possibilities, and by the freedom that also comes with the realization that they are man made. They are more vital for the same reason that they are more variable than professional culture lovers like to think. As for the objective view we want, the method of science is a means of getting above the relativities. It enabled Miss Benedict to make out the patterns that are in any event an important fact in human history. Or a specific illustration at once of the compulsions of culture and the possibility of transcending them is Freud's discovery, so shocking to his contemporaries, of the spontaneous sexual life of children. To a Trobriander or Marquesan, Kardiner observes, this "discovery" is a self-evident fact; yet it was a brilliant insight.

The pertinence of anthropology for esthetics and criticism, finally, is much the same as for the humanities generally. It has also been clouded by the same misconceptions. Thus primitive art has owed much of its recent popularity to mistaken notions of the "primitive." Sophisticates have found it charming as a merely rudimentary expression, like the efforts of children. Others have indiscriminately admired it as "purer" than civilized art—more natural, more fundamental. Doubtless primitive artists are less intellectual and self-conscious than contemporaries; Ralph Linton reports that almost always they are unable to describe their creative processes or define their standards. Doubtless they are more spontaneous, because they and their whole community take for granted that art is as natural as any other activity. Yet their purposes are not single nor their values pure, and both are culturally conditioned.

Thus Herbert Read distinguishes three main purposes in primitive art: hedonistic, or the purely sensational pleasure in ornament; ritualistic, or the symbolic representation of communal ideals; and expressive, or the emotional representation of the artist's own sorrows and delights. Any one purpose is likely to predominate in a given culture, but all are legitimate—and all still persist. Again, MacCurdy declares that the beginnings of art were apparently naturalistic, as in the cave drawings, whereas Otto Rank declares that it generally becomes more abstract as one follows it back; the one thing clear is that both representational and geometrical or conventionalized art are "natural." Like other institutions, moreover, art takes on new meanings and uses as it develops. Its associations in primitive societies would not in any event dictate the kind of cultural company it ought to keep today, but even in these societies its company is varied. The evidence does not support, for instance, the common belief that it always goes hand in hand with religion. In some cultures they are quite independent—as in the pueblos in the Southwest, whose art is of a high order but whose religious objects are of very shoddy workmanship.

In all the cultures we know, men exhibit an esthetic sense, and most likely there are certain fundamental principles of composition and design, just as there are fundamental rhythms and patterns in all human lives. Getting at these roots of art, however, gives no certain knowledge of how it will flower, or warrant for asserting how it ought to flower; the meanings and values of art are also socialized. In this light, then, one may assay such dubious critical counters as "universality" in literature. It is not necessarily meaningless. Granted a decent modesty in such large matters, and a caution against the assumption of something metaphysically anterior or superior, the critic may attribute universality to representations of the basic patterns, the recurrent situations and the recurrent strategies for dealing with them, the needs that seek fulfillment in all cultures, the natural laws that give all behavior continuity and consequence, rewarding and exacting payment from all men in accordance with their capacity for feeling and striving—the underlying logic of experience always and everywhere, because of which the story of a man's rise or fall, or rise and fall, has been told in every language. Thus in his *Philosophy of Literary Form,* Kenneth Burke analyzes literature as a mode of symbolic action; like all social acts, symbolic acts may differ greatly in relevance, scope, and adequacy, but they remain variants of a few fundamental strategies, such as "pious awe (the sublime) and impious rebellion (the ridiculous)." In ordinary usage, however, the "universal" turns out to be a local or even private preference in truth and beauty; the term is a popular device for uttering the unutterable; and anthropology is the most obvious lead to the necessary perspective here.

To be sure, we now take for granted the idea of "climates of opinion," and have made admirable use of it. We are capable of a fuller understanding and finer appreciation of past literature than were our more pious ancestors. If we are more critical we can also be more charitable, perceiving that the great fathers were also children of a country and an age, perceiving that it would be unfair even to Shakespeare to read his plays as if they had been written by a timeless spirit instead of a popular playwright of Elizabethan England. Yet for just this reason we are likely to feel outside of history, talk as if we ourselves were "citizens of eternity" and not merely of the twentieth century. We are too seldom fully aware that our own understandings are also conditioned. Cultural climate is indeed too insidious for the most vigilant, scrupulous critic. It will get him in the mere act of reading: Augustine described his wonder when he saw Ambrose reading without moving his lips, and unquestionably our habit of reading in the mind has influenced our judgments of prose style. Or it will get him in the simplest perceptions: seeing is indeed believing, for what we see is partly an inference or act of belief.*

In any event the past can never be seen or felt as it was lived. Simply because it is the most living thing in the present, it changes with the present; a different Caesar, wrote George H. Mead, is always crossing a different Rubicon. As for literature, *Œdipus Rex,* the *Divine Comedy,* and *Macbeth* cannot possibly have the same meaning for us that they had for their contemporaries, had for the eighteenth century, and will have for the

* In *Technics and Civilization,* Lewis Mumford points out that "no two cultures live conceptually in the same kind of time and space." In the Middle Ages time and space were symbolic, relatively independent systems, their "truest" forms being Heaven and Eternity; the artist was as indifferent to perspective as to anachronism. By 1600, however, they came to be coördinated with movement, accurately observed and represented, measured, *used:* "The conquest of space and time had begun." The conquest has proved stupendous; but now a new story is beginning. While such writers as Proust and Mann have been obsessed with Time, physicists have also been at work on it; the conceptual world of the future bids fair to be different, once Einstein has become common sense. How our world and its products will then look is hard to say, but we can count on looking a little quaint.

twenty-second. For the same reason, however, we cannot afford to be too self-conscious about the temporality of our attitudes. We should not worry too much about the verdict of posterity—posterity will be in the same boat. More to the point, we should not worry too much about the implied verdict of the past. There is a real danger in the wonderful feat of sympathetic reading. Pious literary men saturate themselves in attitudes that have become irrelevant, confuse historical interest with esthetic value, forget that the point in studying the past is to bring its resources to bear on the present—to make it earn its living. As we use the past, in short, so in a real sense we choose it, we make it. We can make the best of it only in a full consciousness that we are both its creatures and its creators.

Altogether, in art as in government, it is idle to provide for human nature in the abstract. We can understand it only in its time and place, we must reckon our own time and place; and for making the best of both past and present, the perspectives of anthropology are invaluable. It is now easy to understand, for instance, the desire of some literary men to shuffle off the coils of civilization, and their call for the good, simple, natural life. Nevertheless the immediate question is *which* simple life; the social behavior of the happiest savage is complex enough, and much more than instinctive. The final answer is that our minds are cultural plants, not containers that can be emptied out like rubbish bins, and that these plants cannot be torn up by the roots and the roots alone then thrust into a different soil. Here is also the answer to the many intellectuals who would bring back old institutions, and who in their nostalgia fail to see the whole configuration either of ancient culture or of their own; the type of modern man who would actually be most at home in Greek civilization, Whitehead observed, is the average professional heavyweight boxer. And if all these relativities are bewildering, at times depressing, it is also exhilarating to contemplate the extraordinarily rich diversity of culture that man has been able to fashion out of the raw materials of existence.

5. A NOTE ON PHILOLOGY AND FOLK LANGUAGE

WE are apt to smile at our ancestors' faith in philology as a means to essential truth. To the Stoics, etymology was the doctrine of true causes (as "etymology" itself indicates); with the development of comparative philology in the nineteenth century, it became a way of getting at the "real" meaning of a word, which would give a pure idea of Reality. Today it is more fashionable to look upon our common language as a forest of petrified fallacy, the chief hindrance to truth seekers. Yet the logical positivists have at bottom much the same faith and the same purpose as the old philologists. They talk of pure symbols instead of eternal essences, they wish to codify the meanings that are sprinkled over the surface instead of enshrined at the heart of reality, they invent instead of discover true causes; but they are simply trying to make language the perfect logical instrument that men had thought it already was, and they regard their own kind of strict meanings as the only kind really worth mentioning. They forget that language serves first and last for the sake of concrete experience, and that in so far as they attain their ideal it even falsifies the actual quality of experience. They forget that the untidiness of common speech reflects the untidiness of experience itself. If the various popular substitutes for the precise term "coitus" have various connotations, Isaac Goldberg remarked, so does the act itself.

At any rate, it is also profitable to study language as an instrument of all social purposes, remembering that when most imperfect it is still the most wonderful of human instruments, the key to the essential difference between human society and all other animal societies, the key to all that is most significant in the tremendous story of civilization. Philology then becomes an especially interesting branch of anthropology and sociology. "If the fragment of a fossil bone can tell us the history of an extinct world," writes A. H. Sayce, "so, too, can the fragment of a word reveal to us the struggles of ancient societies, and ideas and beliefs that have long since perished." More important, philology throws light on ideas and beliefs that have not perished, the whole background of consciousness and thought. In another view, language has been called a bouquet of faded metaphors; to restore the original image can freshen not only our speech but our perceptions, surprising us out of our habitual ways of looking at things, giving the effect of perspective by incongruity. It can thereby restore an awareness of fundamental congruities and continuities, and so help us to escape the provinciality of modernism.

In this spirit we can freely say the worst about our common

language, still emphasize that it reeks with primitive habits of thought. Early peoples stood in awe of mere names, ascribing to words the full significance of things—and then some. Thus primitive man usually cherished his name as an important personal possession, and was often prohibited from mentioning the name of a dead man; Jehovah laid down as the first commandment that His name should not be taken in vain; in Rome not only fame but fate (*fari*) was "that which has been spoken." Similarly the primitive employed euphemisms and circumlocutions to put a psychological distance between himself and fearful realities, and his key words for dealing with these realities give the clue to his mentality; Freud has analyzed the ambivalent attitudes reflected in such words as "taboo," which signified both holy and unclean. This ambivalence is still found in our words having to do with "awful" matters. But it is hardly necessary to add that all such primitive attitudes toward words still persist. Sticks and stones may break your bones but names will never hurt you, chants the youngster. He does not really believe this, and much less does the grown man.

Yet the whole lesson is by no means so simple as this. We incidentally need to be wary in our analogies, for we cannot recapture the entire or exact meaning of primitive symbols. "Soul" or "spirit," for example, does not adequately describe the animistic belief; strictly, we cannot represent or reproduce a scheme of thought which, as anthropologists have shown, made no clear distinction between natural and supernatural, or none at all between logical and illogical, but was geared chiefly to a distinction between the ordinary and the extraordinary. Above all, however, we need to guard against the assumption that all is fallacy here. In *An Inquiry into Meaning and Truth,* Bertrand Russell returns to the study of ordinary language, having rediscovered that it has a significant relation to experience, and that knowledge and truth are born of this relation. He maintains that logical positivists have oversimplified the whole problem precisely because they have ignored these simplicities; they need to begin their analysis on the level of ordinary language, the more primitive perceptions or expressions of experience, instead of on the highly complex level of mathematics and logical syntax. But whatever its epistemological status, our common language is at least rich in the common garden variety of truths. If its roots are sunk in ignorance, superstition, delusion, they are also sunk in the oldest, deepest experience of the race. They reveal the permanencies of interest and desire that are often obscured by all our subtle distinctions and refined analyses. They reveal the natural roots of a naturalistic philosophy.

The earliest metaphors are of course naïvely anthropomorphic, indicating how literally man was the measure of all things. They survive in our most familiar expressions: the leg of a chair, the mouth of a river, the foot of a hill, the heart of a problem. That they have survived also indicates how deeply men have felt at home in their world. Despite the mysterious or hostile powers with which they have peopled it, they have acknowledged in their language that nature is not actually so alien to their interests as philosophers sometimes represent. In our language, the word "interest" itself—*inter-esse*—implies a constant reference to the environment and *inter*action with it. More significant is all the evidence, buried in the metaphorical roots of our abstractions, for the pragmatic view of doing and making as the source and proof of our knowledge. "The testimony of language is clear," writes Goldberg: "the beginning of understanding is in man's physical, sensual experience with the objects that surround him." A "fact" (*facere*) was originally a thing done or made, and all our knowledge "has to do with" the natural environment. To "comprehend" is to grasp or catch on to; to "manage" (*manus*) is to handle or get one's hands on; to "say" (*dicere*) is to point. If we believe that spiritual matters are finally all that matters, they are still "matters," the stuff of this world. Our most elevated like our humblest affairs (*ad-facere*) are "doings to."

To abstract is etymologically to draw from or to separate. Our abstractions carry us so far from their natural source that we are apt to get out of touch with it and then to make sharp divisions. Nevertheless our language points to man's deep desire for unity and wholeness. "Whole," "holy," "hale," "healthy" all come from the same root; "entireness," "integrity," and "catholicity" etymologically express the same idea; to be "satisfied" (*satisfacere*) is to be made whole; and today, when we are "torn" by fear or doubt, we are told to "pull ourselves together." In any event we live in a "universe," a multiplicity "turned into one whole." Thinkers who make a sharp split between nature and human nature, body and soul, are presumably inviting us to holiness, but they are literally unhealthy, unwholesome. Similarly in the realm of social values, where language reflects chiefly the values of the ruling class, patrician distinctions between "noble" and "villein," we can also perceive the deeper ties that unite men. Language itself is possible only because of these ties. To "communicate," to form a "community," is to "put in common." A man can "commune" with himself only because he is a member of a community; to have utter privacy would indeed mean to be "deprived" or bereaved. The source of all morality, the biological basis of the ideals of solidarity, justice, peace, is found in the great metaphor of "mankind"—man as kin. And "kindness" is a natural consequence of the profound identities in kind.

All this leads to the rich resources of folk poetry and folk criticism today, quite apart from the picturesqueness that delights slumming intellectuals. No doubt popular speech is often overrated. Its famed economy is apt to be the economy of the rubber-stamp, its simplicity the simplicity of the schoolboy, its force a forcible-feebleness; its poetry is often trite metaphor or doggerel, its criticism a cocky snap judgment or the mere mulishness of horse sense; and today especially it is often corrupted by the jargon of science and industry, the verbal vices of the intelligentsia. Nevertheless there remains a rich vein, particularly of folk criticism, that the educated or cultured class turn away from because of a genteel fear of being common or a bourgeois fear of being incorrect—much the same sentiments exploited by Emily Post.

You can take it from the common man because he is always likely to be put on the spot or left holding the bag, can never get away with much. He does not have to be very sensitive to be often shrewder in his judgments, more accurate in his namings, than are his intellectual betters; if he is too often taken in, he can also spot the high-brow, the stuffed shirt, the windbag, the fourflusher, the chiseler—the kind of fake that intellectuals may be quick to detect in industry and politics but are likely to be fooled by in the world of art and thought. Furthermore, in the language of the common man live many poor relations of aristocratic words, useful in reminding us of the common ancestry and toning down pretensions. "Classic" is also "classy"—and "classy" is the more accurate word for much art of the eighteenth century, and for some critical attitudes today. Above all, in its hospitality popular speech admits important truths that are forgotten or excluded by the experts in truth. Thus Dewey has pointed out that the idiomatic use of "mind" is more scientific and philosophical than most technical use. We keep things in mind and call them to mind, we bring mind to bear, we have a mind to do something, we change our mind, we mind our step, we mind our parents, we mind our babies, we mind if we are insulted—in our everyday life we recognize that mind is an instrument for all practical, emotional, volitional as well as "intellectual" purposes, and is involved in all our actions. Philosophers who regard it as pure reason, divorce it from practice, are simply absent-minded.

Even the fallacies engrained in primitive and folk language may contain an element of natural wisdom. The subject-predicate proposition has doubtless had seriously unfortunate effects upon thought; the most dangerously ambiguous word in the language is still the simple "to be." (Equivalence, identity, definition, predication, existence, actuality, derivation—Santayana has shown that these are among the meanings of "is" that are constantly confused.) Yet it remains not only an indispensable but a "realistic" word, useful for disciplining the imagination of scientists as well as poets. It is the more necessary when scientists and philosophers are prone to talk only of fictions, symbols, pointer-readings, thought stuff. The first and last answer to Eddington as to Berkeley remains Dr. Johnson's kick: the stone *is* and it is *there*. We might try to restore the noun "being" to its original status as a participle; meanwhile the stone is a solid thing as well as a rarefied electronic event, with fixed "properties" as important practically as dubious philosophically, a reminder of the permanencies in a sea of change. And in other ways thinkers would do well to recall at times their aboriginal innocence. Primitive language makes plain that men are quicker in seeing the differences than the similarities in things. Thus the Zulus had different words for different kinds of cow but no word for cow; they had not reached the abstract idea of cow-ness. Today cow-ness is an immediate perception, the point of departure for a train of higher abstractions going from species to genus to animal kingdom to living organism to electric charges moving in a field of force—to no man's land. The substance of our knowledge lies in this perception of ever more inclusive similarities. Nevertheless we should not forget the cows. It is now time, as

Korzybski insists, to stress again the *differences*, the particularities left out of our abstractions.

All such suggestions, of course, are merely suggestive. Our common language is also rich in less agreeable ones: the ties between economics and ethics indicated by bonus, good and goods; the grosser materialism indicated by "good as gold," "a heart of gold." If the whole history of society is summarized in language, it has not been simply an edifying story. Nevertheless it is our story; and this kind of study leads us back to a "commonwealth" in the literal, ancient sense of the word.

Another context for philology, finally, is suggested by evolutionary theory. Nobody can know how or when men began to organize their grunts; we can merely speculate on the origins of language—as in the "bowwow," "pooh-pooh," "dingdong" theories. We take for granted, however, that it grew much like society itself, unconsciously, organically, and we can see that its development followed the general course of evolution, from simplicity to patterned complexity, from automatic adaptation to conscious control. Hence we may reasonably assume that the general principles governing its structure and function should correspond more or less with the general laws of biology, psychology, and sociology. E. E. Southard, for instance, has pointed to a resemblance between the four grammatical modes and the four "humors": imperative—choleric, indicative—phlegmatic, subjunctive—melancholic, optative—sanguine. He believes that there is a necessary, close correspondence between the possible forms of linguistic structure and the basic structures of personality. But a still wider correlation is suggested by George K. Zipf in *The Psycho-Biology of Language,* an introduction to "dynamic philology."

Zipf analyzes "the structure and forces of configurational arrangement" in language, on the assumption that patterns of speech might be expected to be commensurable with other patterns of behavior, and he finds considerable underlying orderliness and uniformity in the phonetic development of all languages. The available evidence points to two fundamental conditions of this development:

1) whether viewed as a whole or in part, the form of all speech-elements or speech-patterns is intimately associated with their behavior, the one changing with the other, so that all seems to be

relative and nothing absolute in linguistic expression; and 2) all speech-elements or language-patterns are impelled and directed in their behavior by a fundamental law of economy in which is the desire to maintain an equilibrium between form and behavior.

Specifically, Zipf finds a close relation between frequency of usage and length and accent of words, the most common words tending to be shorter, the most common syllables or elements to be unaccented; and endings and declensions appear to be similarly determined by the "dynamics of the human organism."

Here again we run into the key words of modern science: dynamics, organism, configuration, pattern. It would accordingly be pleasant to report some illuminating discoveries and exciting prospects. Actually, however, the findings of "dynamic philology" are as yet pretty meager and tentative, and even the possibilities suggested by Zipf seem to me unexciting. Yet he points to organic connections worth keeping in mind, even if they do not bear significantly on the chief values of language. His theory is useful at least as a frame or background for other, more obviously fruitful researches.

6. POSTSCRIPT: THE INDIVIDUAL AND SOCIETY

The State, Thomas Hobbes declared, was the sole means of enforcing the social covenant that raised men above the condition of Nature, the condition of "a Warre of every man against every man," and the life of all "solitary, poore, nasty, brutish, and short." On the other hand, eighteenth-century thinkers with more agreeable ideas about human nature were apt to have more disagreeable ideas about the State. Implicit in almost all these diverse social theories, however, was an atomic individualism that provided no more inherent ties of association than did classical physics; society was held together only by enlightened self-interest—with the policeman on the corner to keep an eye on the unenlightened. Today sociologists are generally agreed that individuals did not form a community but emerged as individuals from a community, and in no extremity of revolt can sever their organic ties with it; their consciousness, the very self they prize, is a social product. Yet most social thinking still follows, if unconsciously, the old tradition. Men usually assume a necessary antagonism between society and the individual, as if each has a separate sphere and what one gains the other must lose. This assumption is apparent in Pareto, for example: he constantly stresses the incompatibilities between social utility and individual desire, ridicules arguments for the greatest good of the greatest number. The assumption is balder in the familiar antitheses —individualism versus collectivism, freedom versus authority. It is indeed understandable: since men naturally think of themselves as uncaused causes, and of all the contents of their minds as their very own, the world in which they have their being seems separate, something else again, a potential threat to their being. And so the vital issues that are at stake today are constantly confused.

Now unquestionably there is some real conflict here, never a perfect community of interest. But it is by no means so evident that the opponents are "the individual" and "society." So conceived, these terms are fictions, and they adequately represent neither party. The individual is thought of as the sum or essence of his differences, his distinctive personal traits, and therefore as naturally asocial if not antisocial; society is identified with a particular group in authority, a bureaucracy. Actually, the concrete historical conflicts (beyond merely private squabbles) have always been between groups within society, the individual acting as the representative or the leader of a group. Such conflicts are inevitable, I dare say, even in a classless society. There will always be divergent opinions and clashing interests, if not a competition for power or prestige; there will always be some kind of government (at least so long as there is a civilization), which will not adequately serve all the interests of all its citizens. Nevertheless these constant stresses or distresses are far from implying a necessary, constant incompatibility between the desires of the individual and the welfare of the whole group. On the face of it, coördination and coöperation are at least as apparent as division and opposition. On second thought, one can hardly deny that only in a highly developed social order can there be opportunity for a full development of individuality. The social bond is a deep, broad, permanent community of interest, a marriage that is not a contract or even a sacrament but a strictly indissoluble union.

Yet the organic concept also has its extremes, which in turn obscure the concrete social reality. Some thinkers appear to conceive a social whole that is not only more than the sum of its parts but practically independent of them. The individual becomes a *mere* fiction, inconvenient or even unnecessary for the purposes of thought—A. W. Small referred to him as a "discredited hypothesis." My immediate objection to this way of thinking no doubt springs from humanistic sentiment; not to mention my feeling that I am not a fiction, and my desire to become a creditable hypothesis on my own terms. But there are solid logical and scientific grounds for objection. There can be no social reality independent of the existence and behavior of individuals, no social well-being unless it is experienced by them; they are the final referent for all valid abstractions. The anthropologist studies a whole culture, whose carriers seldom have a clear understanding or even awareness of it; nevertheless they alone carry it, and they transmit it in specific enough ways. The sociologist goes behind the individual to study social consciousness and formulate the laws of social behavior; nevertheless he can directly know only the experiences of individuals, really observe only their habits of behavior. In short, we can no more understand society apart from them than we can understand them apart from society. Historical processes are never in fact wholly impersonal; they are knowable only as the deeds or misdeeds of men. And great men have played a greater part in them. If this is sentiment again, it is still hard to refute Mill's statement that all wise and noble things have been not only personified but initiated by individuals.

Today there is especial reason to reassert the claims of the individual. Whatever their theoretical implications, the actual tendency of science and industry, and of such derivative creeds as Marxism, has been primarily quantitative, mechanistic, necessitarian. The individual is buried in the statistical average, the machine standard, the impersonal law. Science, industry, the classless society are all presumably designed to promote the welfare of this nameless creature; "what helps business helps you," thanks to science you can now expect to live longer, come the revolution and you'll be free; but you become a nuisance if you express doubts or ask "meaningless" questions, and you must wonder why so much trouble should be taken to provide a bigger and better denominator for so insignificant a numerator. If these issues still seem somewhat theoretical, they are certainly practical and urgent enough in the political sphere. Here we can never afford to minimize the perpetual tendencies to oppress, stifle, persecute, enslave—the tyranny of majorities and minorities alike. The tragic dignity of history is summed up in precisely this endless struggle for human liberties. And today the political equivalent of the extreme organic view that discredits the individual is totalitarianism, the worship of the State apart from the welfare of its members.

This very concern for the individual, however, demands that the final stress fall upon his organic relation to society. Only as a social being does he have conscious rights, interests, a mind of his own. Least of all as we respect him can we regard him as actually or ideally self-sufficient, a law unto himself. The most solitary hermit carries into his cave a mind formed in his society; we may even suspect, Cooley remarked, that St. Simeon Stylites was not unaware that his austerity was visible to others. The fiercest rebel against society has accepted the bulk of its culture, obeys most of its conventions, and is indebted to it for the very principles in the name of which he rebels. The man who most nearly cuts all bonds is not the hermit, the outlaw, or the anarchist, but the lunatic. "We belong to a group," writes Mannheim, "not only because we are born into it, not merely because we profess to belong to it, nor finally because we give it our loyalty and allegiance, but primarily because we see the world and certain things in the world the way it does." We get from it even our elementary meanings and emotions—without it we would have only sensations. We get from it, in the strictest sense, our very selves. The self is not significant until it has become aware of itself, capable of what George H. Mead called the "reflexive mood," in which it becomes both subject and object of an experience. Self-consciousness is achieved through social intercourse, by adaptation to others and not merely by a narcissistic contemplation. In Mead's terms, the self becomes aware only as it plays the roles of others, gets outside itself—becomes itself a society.

The richest self, to restate the ideal of humanism in these terms, would then be the widest society, the self that can play the most roles with the most ease and skill—the self that is least selfish or self-centered. The measure of maturity, of individuality in the deepest sense, is the range of our sympathetic understanding. In any event, utter egoism is as inconceivable as utter selflessness. The self has its existence within the general life, not outside it, and can never be absolutely separated from other selves. The "I" includes "you" and "they," becomes "we"; empty "my" mind of all thoughts of others, all ideas of the general life, and I cannot imagine what would be left; and if I am jealous of my possessions, material and immaterial, these include my duties, my ideals, my share in the common cause. Even private thought, as has often been remarked, is a conversation with oneself—a "communing"—in which the implied listener is the group. It supposes at least an imaginary communication, but it receives its natural fulfillment when it is expressed, made public. Self-expression is literally necessary for self-preservation; for without communication there can be no thought, no self.

Likewise there is no absolute antithesis between egoism and altruism in the realm of morals or social behavior. The most selfish person is still seeking objects of common desire—and with them the approval of others; what we usually call selfishness is less fundamental and less universal than this desire for social approval. Nor is fellow feeling or altruism merely a cultivated sentiment. Rights are universal, Rousseau declared, because logically no man can claim them without recognizing that others must have them too; more important than this logic is the fact that a total disregard of the rights of others is literally impossible. Some kind of moral sense is the very condition of consciousness, some kind of moral law is implicit in the very fact of social life. The policeman on the corner is less the guardian than the symbol of the social order. Thus Piaget has shown that the very young child is indeed egocentric, in his conversation and his games having chiefly an illusion of give and take, and that his obedience is a blind, mechanical submission to incomprehensible rules. As he grows up, however, he becomes really aware of others, and with genuine social relations comes a personal understanding of morality and justice; obedience to rules is then an inner necessity. Needless to add, he sometimes manages to disregard this necessity. He grows up to become more or less of a problem to his fellows, and to himself. But the point remains, contrary to a fairly general impression, that he is not by nature a complete egoist who has to be broken into social behavior against all his instincts, as an animal is trained to the yoke; that he achieves his prized selfhood primarily through what unites him to his fellows, not what distinguishes him; that he achieves freedom through social consciousness, not in spite of it. The recognition of this natural union between the individual and the group is a far sounder basis for efforts to minimize and adjust the natural conflicts, to accommodate the diverse needs of diverse individuals, than is the assumption of grudging compromises or temporary truces between irreconcilable antagonists.

In modern society, to be sure, "community of interest" may seem a fiction or mere metaphor. Apart from the violent antagonisms between nations and within nations, men have become sharply distinguished and often separated by a minute division of labor, an extreme heterogeneity of interest, a rapid growth of uncommon meanings. All that has so greatly increased their actual dependence upon one another has also obscured it. Hence solidarity becomes an intellectual concept, an ideal to be preached; community becomes commun*ism*. Nevertheless interdependence remains the fundamental fact in our civilization, the necessary basis for all efforts to deal with its problems. At the same time—and this is most important—interdependence does not necessarily mean a loss of independence. The growth of civilization has manifestly brought new possibilities of self-fulfillment, new forms of freedom. With the possibilities come new responsibilities, and with the freedom new compulsions. We cannot have one without the other; a complex industrial society cannot also have the advantages of the simple life. The privileges men come to take for granted, however; and then they are conscious chiefly of the compulsions. Thus they complain of regimentation—forgetting that actually they are freer to do many more things than were their unregimented ancestors. Thus, in other moods, they envy the carefree child—who is full of cares, and whose main grievance is that he is not at all free.

For the artist, this whole issue is especially acute today. He naturally feels very much an individual, often very superior to the group. He cherishes the individuality that has been the characteristic goal of gifted men ever since the Renaissance, on lower levels becoming the "personality" that simple citizens try to achieve with the aid of an illustrated booklet. He has probably been affected by the romantic cult of genius, the religion of art —which on lower levels, again, appears as the exploitation of the "artistic temperament," the feeling of the most trivial Bohemian that society owes him a living as a mere token payment on the incalculable debt owed by posterity. But often, too, the artist feels very lonely, an outsider or even an outcast. The increasing specialization in modern civilization, the terrific competition of science and technology, the rise to power of a moralistic and materialistic bourgeoisie, the extraordinary growth of gadgetry —such conditions have tended to isolate him in fact. He who was once the inspired voice of the tribe, and then the inspired "maker," no longer feels at the center of the vital processes of doing and making. And because he is often out of sympathy with the insensitive masses, or rebels against particular constraints, he is apt to be pictured as an eternal alien or eternal rebel. Herbert Read has asserted that "the poet is necessarily an anarchist," the agent of destruction for all social forms and traditions, which are in themselves "attributes of death."

Now we need not worry that the poet will in fact become an utter anarchist. He cannot write a word without acknowledging his debt to society for his consciousness and his medium; form and tradition are as essential for art as for society because implicit in the living consciousness, the very structure of the organism. "Artists never stand by themselves," Nietzsche declared, "standing alone is opposed to their deepest instincts." In our time the most anarchical were not content to ride their dadas but kept issuing manifestoes, became Dadaists. But so much energy and talent have been squandered in these confusions, from symbolism to surrealism, that we need to return to first principles, and specifically the principle of the included middle. Individualism and collectivism, freedom and tradition, self-expression and communication, romanticism and classicism—these, once more, are not antitheses, and both terms refer to tendencies at work in all artists at all times. In a given artist at a given time one or the other may dominate; there is always some conflict, artistry is always a discipline. But one does not necessarily gain at the expense of the other, and often enough the artist approaches the obvious ideal of balance and harmony. If the critic must dwell on one or the other as he places or describes, the more as he tries to combat a given excess, he need not plant his standard for all time at either.

There is always occasion to repeat that the artist is conspicuously an individual, speaking for himself as well as for the community, speaking because he has something of his own to give to the community. We are apt to minimize or even miss the individuality in art work of earlier times or other countries, for much the same reason that all Chinamen look alike to us; in the relatively remote or unfamiliar we see chiefly the type. In particular, primitive and classic art alike have supported the idea that ideally the artist is only the voice of the tribe; the many variations on Plato's theory that he is one "inspired and possessed" represent him as the voice of deity, or the medium of some impersonal revelation; the less exalted theory of "mimesis" also tends to depersonalize him in the business of holding up a mirror to nature; the modern preoccupation with underground forces leads to an emphasis upon his work solely as a product of the age, the class, or the id; and critics are always prone to abstract a few generic traits and ignore the individual quality of his work, precisely as physicists ignore the color, feel, and emotional significance of things in order to measure and weigh them. Yet the most rigorous conventions or the most reverent effort at conformity cannot utterly suppress his individuality—the distinctive style that is indeed the man. If he often makes mechanical use of custom-made materials, he contributes something of his own the moment he becomes "creative." As he approaches greatness he is more profoundly original, whether conformist or nonconformist. The universal genius is the most striking proof of uniqueness, and of all that the individual can contribute.

At the same time, genius cannot be expressed or achieved outside a society, and the richness of its fulfillment is the more dependent upon the richness of its cultural heritage; Homer was produced by Greece, he could never have been a Homer in Dobu. Today the self-conscious artist may forget that he always expresses far more than himself—and that the only "pure" self-expression is a twitch or a howl. From its inception, his experience is influenced by the age-old habit of communication. When he begins to express or objectify it he further socializes it; his own satisfaction lies in achieving the right "effect," which is ultimately the effect upon other people. His artistic conscience is also the voice of his public, however limited in numbers or cultish in taste. In short, because the artist is a self-conscious individual he is especially fitted to be the spokesman of the group. For him especially is the ideal stated by Kierkegaard:

> The individual becomes conscious of himself as being this particular individual with particular gifts, tendencies, impulses, passions, under the influence of a particular environment, as a particular product of his milieu. He who becomes thus conscious of himself assumes all this as part of his own responsibility. At the moment of choice he is thus in complete isolation, for he withdraws from his surroundings; and yet he is in complete continuity, for he chooses himself as prod-

uct; and this choice is a free choice, so that we might even say, when he chooses himself as product, that he is producing himself.

Today the milieu is a welter, continuity an immense tangle, responsibility a choice from multifarious possibilities. When the individual artist has produced himself under such difficulties, he is apt to feel responsible only to himself, and to himself only as an artist. If the poet is man at his most self-conscious, Louis MacNeice writes, he should be conscious of himself as a *man,* not merely as a poet; but he now feels obliged to take up a *peculiarly* poetic position. Hence he has cultivated his private sensibility, or written exclusively for other poets. Often he has shied away from all the intellectual, moral, social, "unpoetic" uses of poetry that the old writers took in their stride because they wrote as men and for the community. And so we return to our central problem: the monopolization of our spiritual capital, the isolation of our spiritual agencies from the communal life. The poet's problem is not to be met, of course, by a mere statement of his actual indebtedness to society and his ideal responsibility to it; he has still to create an audience as well as a work even when, like Auden, he feels an urgent social responsibility. But a clear perception of the actual ties between the self and society, the nature of community and communication, is at least a beginning toward more satisfactory public relations. As it is, too many poets and critics have aggravated the problem by a one-sided view of it. They have sacrificed culture for the sake of cult, integration for a narrow integrity, wholeness for an artificial holiness. In various ways they have made of their isolation a policy of isolationism.

PART III

RESTATEMENT

VIII

THE PURPORT OF SCIENTIFIC HUMANISM

For the matter in hand is no mere felicity of speculation, but the real business and fortunes of the human race, and all power of operation. For man is but the servant and interpreter of nature: what he does and what he knows is only what he has observed of nature's order in fact or in thought; beyond this he knows nothing and can do nothing.

FRANCIS BACON.

1. THE RELATION TO PHILOSOPHY

IN the account of the world given by nineteenth-century science, a machine was a much more satisfactory thing than a human being. It obeyed all the laws, it had no "subjective" nonsense about it, it completely satisfied the definition of "reality"; the living consciousness of man was simply a nuisance, a constant hazard for truth seekers even after it had been explained away as an illusion. Today, as we have seen, the leaders of science no longer pretend that it gives us reality, the whole reality, and nothing but reality. They acknowledge that it is a selection from reality, a reconstruction for special purposes, and that as a human interpretation it too falls back on something like the pathetic fallacy—although, as in good poetry, this is never merely pathetic or simply a fallacy. They have put the machine back in its place, taken the curse off consciousness, left room for other selections and other purposes.

More specifically, to recapitulate, the trend in scientific thought is indicated by the constant recurrence of such terms as "continuity," "evolution," "interrelation," "integration," "system," "field," "pattern"—all summed up in the concept of dynamic, organic wholes. No doubt these words are suffering from overwork, and may lead to easy verbal solutions. Thinkers may use the symbol "energy," for instance, to harmonize and unify everything under the sun; for everything can be viewed as a form of energy. Nevertheless the new concepts are not merely new sayings. They work, they enable more effective doings. At any rate they are a significant means to harmony and unity. They bring scientific thought closer to concrete experience and vital purpose, the "whole situation" that ordinary practical thought tries to grasp. They reclaim many affirmations of idealists, intuitionists, vitalists, pantheists, or mystics at a higher level, where experimental inquiry and rational analysis are also possible. They enable, I believe, the most coherent, comprehensive account of our knowledge and experience.

And so I return to my beginnings, to restate the philosophy that may now be given the current name of "scientific humanism." Science grew out of the ancient tradition of humanism, with its faith in light as the primary means to the good life on earth, and has in turn substantiated and amplified this tradition. It has stressed the need of tolerance, flexibility, and catholicity by stressing the concept of relativity and the reference system, the attitudes of "as if" and "that depends," the admission of plural possibilities. It has contributed notably the idea of the extraordinary plasticity of man, the almost limitless possibilities of nurture implicit in nature, and the idea of an evolving, unfinished universe in which man can play a creative role, take his own purposes seriously even if the universe doesn't. Science also gives a cue for this role in its experimental logic, the adventurous but critical attitude that is called for by the assumption of potentialities and the admission of ultimate uncertainty. It is in general an impressive demonstration of humanistic values in practice. Despite its supposed inhumanity—and the actual tendency of many disciples to strut too much and fret too little—science exemplifies the ideal mean between pride and humility, faith and doubt, freedom and obedience, individual initiative and collective discipline. It is the most striking evidence, once more, of the real possibility of "supra-personal, supra-partisan, supra-racial standards and values."

If we reject the absolutes of Newton as of Aquinas, we perforce talk in metaphor. There is no strictly scientific procedure by which to choose among the comprehensive metaphors—the faiths and the philosophies that body out the immense metaphor Reality. Mortimer Adler, to be sure, states as an axiom that properly there are no systems of philosophy: there is philosophical knowledge that is more or less adequately possessed by different men, that is absolutely true, and that is superior to science because it is a knowledge not of phenomenal manifestations but of the being of things, sensible and suprasensible. Nevertheless such axioms simply make all argument futile. One may object that Adler cannot possibly prove his statement and that the entire history of philosophy disproves it; in the end one can only say that it just isn't so. If discussion is to be at all fruitful, we must fall back on something like the test suggested by Kenneth Burke. Let each man fill out his metaphor, apply it concretely, show its range and relevancy, say and do all he can with it. This is what even the absolutist tries to do; this is the test that his thought has to meet anyway.

Most ethical philosophers have agreed that the ultimate good, of which pleasure is only a sign or by-product, is the highest and most harmonious development of man's powers, the fullest realization of his capacity for knowing, feeling, and striving. The source of this ideal is the elemental drive that can be made out in all cultures: the desire for a feeling of effectiveness. This accounts for the intrinsic gratification of work and underlies such specific motives as the "aggressive instinct" or the creative impulse. It explains, according to Abram Kardiner, the destructiveness and the pleasure in cruelty that have been attributed to regression, a sadistic instinct, a death instinct, etc.; such behavior is an effort to express mastery when natural channels have been blocked or natural techniques not yet learned. The immediate argument for scientific humanism, then, is that it makes full use of the most effective technique yet devised to master the environment; in Burke's terms, it can tell us most accurately what to look for and what to look out for. But in general we can say more by it and do more with it. Instinctive man, behaviorist man, economic man, spiritual man, innately good or evil man—such metaphors are all relevant but all limited. The philosophies founded upon them too narrowly define the natural possibilities of self-fulfillment, slighting either our knowings or our feelings. None gives full scope to the plasticity, expansiveness, and resourcefulness of man, the elastic measure of an evolving world.

Now this whole study has been an effort to show the range and relevancy of scientific humanism, particularly through its relation to science. More helpful in summing up, however, is the relation to philosophy—and Adler notwithstanding, to certain systems of philosophy. Although the common man regards philosophers as useless, if very deep fellows, they have had a profound influence on the body of meanings that govern social action, the ideas that the common man lives by. Nothing is more important than philosophy in this broad sense; it involves such consequences as world depressions and world wars, which should be practical enough for any man. Formally defined, it is still indispensable as the only discipline that systematically criticizes first and last principles, the ends of all thought and action. Hence it is a necessary supplement to science, both because science is a severely limited inquiry and because even physics presupposes a metaphysic. Philosophers can offer a critique of the assumptions of science, an analysis of its methods, a synthesis of its findings. Historically, indeed, they have tried to establish some special reality or truth of their own, apart from science, and they have tended to divorce theory and practice; but this very effort has had important practical consequences.

"Its Centre is everywhere, but its circumference nowhere," exclaimed Joseph Glanvill as he listed the delights of philosophy, the "easie twinkle of an Intellectual Eye." Already, however, his contemporaries were fixing on a center, and they were seldom twinklers. As science took over the objective world, philosophers turned to the subject, the reasoning self; *cogito ergo sum* signaled the breakup of the unified world view of the Middle Ages. In this way philosophers prepared the way for the rise of psychology, but meanwhile they had the mind to themselves. In this way, too, they raised the problem of how the cogitating self could know anything about the external world. Hence they fought out on epistemological grounds what common men were experiencing as religious conflict. They soundly criticized the basic assumptions of science; Berkeley, for example, demonstrated that if the "secondary" qualities of matter are "unreal," so are the "primary," for ultimately both are "subjective." But they also effected the sharp split that became the main source of philosophical confusion. They ended by making it logically impossible for us to get in touch with the objective world, and by leaving the mind somehow to get along by itself.

Santayana has concisely summarized both the valuable services and the serious disservices of the famous critics of reason—Berkeley, Hume, Kant. They exposed the natural habits and necessary fictions of human thought, its "tentative, practical, and hypothetical nature"; they showed that all our empirical knowledge is based on inference. Then, however, they unnecessarily concluded that the inference is untrustworthy, the knowledge groundless, and they offered in its stead some purely arbitrary faith. Although Hume brought in no *deus ex machina,* he failed to see that he had left no logical foundation for his interests as a philosophical man of the world; but Berkeley's skepticism was only a means to an unutterable, unaccountable kind of truth, and as for Kant, there is a rather incredible effrontery in the shadowy imperatives that he substituted, in the name of "practical reason," for the solid structure of empirical understanding. All were obsessed by the fact that we are men thinking, an inexplicable fact but still an elementary one. All tended to forget the as elementary fact that by thinking we have learned a great deal about the world, and know far more than our epistemologically innocent ancestors.

Let us grant, at once and for all, that the contradictions in formal philosophy pointed out by the early Greek skeptics cannot be utterly resolved; our understandings must always be vulnerable to skeptics because we can explain neither the data nor the processes of inference. Nevertheless we may transcend the contradictions by the simple observation of how men actually behave amid epistemological puzzles; as Engels said, "Human action had solved the difficulty long before human ingenuity invented it." Thus psychologists have discovered that patients suffering from "psychical blindness"—an inability to recognize what they see, to see a thing for what it is—can manage remarkably well by the aid of inferences. On the street they can distinguish men from vehicles: men are long and narrow, vehicles are wide. Call our knowledge "mere inference" and it is valid in the same way; it saves us from getting run over in life. If we can draw no sharp line between relative and absolute truth, we can at least make clear enough distinctions for all practical purposes, asserting confidently that the biologist's cow is more real than the cow that jumped over the moon. In any event we do not simply make up truth in our heads or have it altogether our own way. We speak our minds, we speak for ourselves, ultimately we speak in metaphor; but we can speak to the point, and the point is also determined by something outside our heads.

Since Kant, philosophers have accordingly worked to bridge the gulf between the objective world and the cogitating mind, and more specifically between science and philosophy. They have worked in various ways, and some have returned to older ways; but the main tendency over the last hundred years has been to make philosophy more empirical and science more philosophical. It is now possible to have a clearer, cooler realization of at once the essential limits and the full relevancy of natural knowledge. Through this realization lies the approach to scientific humanism, as I conceive it. And to locate my position somewhat more precisely, to chart the currents of thought leading to it and away from it, I shall briefly consider three representative philosophers—Henri Bergson, William James, and John Dewey.

Although Bergson and James evolved their thought independently, they had a common cause: the attack on intellectualism. Both wished to make philosophy more genuinely empirical by bringing it closer to the immediately given, accepting the world at its face value. They were unwilling, Ralph Barton Perry writes, "to deny immediacy the title of knowledge, because, although it can reveal only the flowing and qualitative aspect of things, this *is* an aspect of things, to which there is no mode of access save by immediacy." Both had a sense "of the copiousness of reality, and of the pathetic thinness of the concepts with which the human mind endeavors to represent it"; for both the central problem was therefore "to reconcile the partial truth of conceptual knowledge with the fuller truth of immediacy." In simpler words, they wanted to bring philosophy back to life.

This general aim should commend itself to humanists. Bergson in particular brilliantly analyzed "the misuse of mind," the habits of verbal description and logical analysis that have led thinkers to confuse their fictions with empirical facts. At the same time, both men were on the forefront of scientific thought. As Samuel Alexander observed, Bergson was the first modern philosopher to "take time seriously," and thereby to get completely away from the static materialism of the past. He anticipated the breakup of absolute space and time in physics; he developed his famous theory of creative evolution, in which he pictured a constant dynamic interpenetration of matter and mind or "memory"; he stressed the unconscious memories that psychoanalysis was to explore, as well as the continuous stream of consciousness; he had, generally, an organic conception of the life process. Similarly James was in the main line of advance in his attack on atomistic and mechanistic concepts, his concern with what something does instead of what it is, his conception of reality as a creation and of truth as a formula for action, his emphasis—more particularly in relation to the sciences of man—upon the total concrete situation. So far, there is an evident reconciliation of science and philosophy, and of both with immediate experience.

Bergson, however, went much farther. He ended at the opposite extreme, where conceptual knowledge turns out to be no knowledge at all. He asserted that the business of science is to *explain* reality, the business of philosophy is to *know* it; we know by direct intuition and not by the use of intellect, which merely explains, even hinders knowing; and what we really know is precisely the thing-in-itself, which Kant naturally could never reach by intellectual analysis. Since this intuition is essentially an esthetic mode of apprehension, Bergson's philosophy is attractive to literary men. He constantly appeals to the arts to support it; he confirms the cherished idea of art lovers, that art is a grasp of the essential truth. Yet just here arises the critical question about this philosophy. If the artist knows reality and the scientist explains it, what is left for the philosopher to do? If he is to give up intellectual analysis, how does his business differ from the poet's? How can he, except in poetry, talk about the truth he knows?

The answer is that he can't. Bergson's picture of the essential reality is not only very vague but self-contradictory. Karin Stephen writes an appreciative book to show why he must be self-contradictory if his theory is true. He has to describe in abstract terms what is indescribable in these terms, he has to use some concepts (such as "intuition") when any concept obscures the actual situation, he has constantly to explain why explanations merely explain or do not really explain. "Plurality" and "unity," for example, are abstractions; the facts of experience are not plural but neither are they (it) singular, since there is only "duration" or going-onness; and at any given point or instant Bergson must say that there really are no points or instants. Mrs. Stephen does not appear to think that these inescapable contradictions invalidate his philosophy. To me, however, they suggest the inevitable futility of any radical attack by intellect upon intellect; I cannot help picturing a blind dog incessantly chasing a tail it doesn't have. At best, this philosophy must exhaust itself in a roll of the eyes, a deep breath, a cry of ecstasy. If it is indeed true, the rest should be silence. There is nothing more that the philosopher can say about it or do with it; he can only repeat or deny; and the more he argues, the farther he gets from his true reality.

But we have no way of knowing whether it is true, no way of being sure that our intuitions are a marriage with reality and have actually won its heart. We have to take Bergson's philosophy on faith; he provides no authority for it, and in the nature of his case can provide none. If intellectual analysis is unnecessary for philosophy, and even a hindrance to actual knowing, children or illiterates ought to be truer philosophers than we; but we cannot seriously believe this. The only possible test of the validity of intuitive knowledge is some kind of rational test. And this is the test we necessarily apply in our serious business with reality. We critically examine our intuitions, we check them against our knowledge and past experience, so far as possible we verify them. In any useful sense of the word, "intuition" itself needs to be known, and it can be known only by conscious intelligence. Conceived as immediate perception, it is essential to knowledge as the source of the data; but if reasoning may then be described as a "sequence of linked intuitions" (Olaf Stapledon's phrase), the ordering is all important. Conceived as an instinctive knowing, a kind of organic hunch, it is often more adequate than conscious knowing; but it is still not trustworthy, and there is no good reason for viewing a more rudimentary form of knowing as the "highest." Conceived as insight, the first flashing vision of all new ideas, it tells us where to look and what to look for; but only intellect can tell us what we have found.*

This is also to say that we cannot in fact sharply separate intuition and reason. More refined methods of analysis can also give access to "the flowing and qualitative aspect of things," and embrace all that is valid in Bergson's intuitive knowledge, stopping short only at the question—which is strictly unanswerable

* Proust, who gave Bergson's thought its supreme literary embodiment, is a striking witness here. The intention of *Remembrance of Things Past* is contained in these words of Bergson: "Our whole inner life is like a single sentence, begun from the first awakening of consciousness, but nowhere broken by a full stop. And so I think that our whole past is there, subconscious—I mean present to us in such a way that our consciousness, to become aware of it, need not go outside itself nor add anything foreign: to perceive clearly all that it contains, or rather all that it is, it has only to put aside an obstacle, to lift a veil." "Any present bare sensation," Karin Stephen adds, "itself suffices to recall, in some sense, the whole past." Thus the famous madeleine led Proust to recapture his whole past. Yet Proust *knew* his past, recaptured its essence, only *after* an immense labor of intellectual analysis. Although he believed that in sensory impressions lay the essential truth, he thought of them as like a snapshot film that must be brought under a light and developed: "You cannot tell what you have taken until you have submitted it to your intelligence."

anyway—of whether this is knowledge of the thing-in-itself. His theory of creative evolution is included in the organismic view; all that is lost is the tautology of the *élan vital.* Similarly "dynamic logic" or "organic thinking" can handle the problems of continuous process and interrelation that he gave over to intuition. "Duration," he held, could not be described in logical terms, for "unity" would exclude "change"; but here he was identifying logic with Aristotelian logic. In short, all that is most valuable in his thought can be and is retained in scientific humanism.

Unfortunately, however, all that is most dubious in it is also still alive. Although Bergson often represented intuition as a complement to reason, he as often separated and opposed them, and as the latter conception was the more striking, his intuition has become popular chiefly as a mysterious faculty that authorizes belief in what reason denies. Similarly his *élan vital* has become popular as a mysterious instinct or creative force superior to intelligence; and a blind creative force can be blindly destructive. He has accordingly contributed to the dangerous reaction against reason that has led the mind to distrust its own powers, deny its own dignity, accept not humility but humiliation.* Let us repeat: today we cannot accept Mill's fiction of the Reasoning Man. We know that reason has an irrational basis, that conscious thought has unconscious motives, that disinterested thought is conditioned by cultural if not class interests, that objective thought is limited by its object. Nevertheless the fiction is invoked even by those who scorn it; in the act of argument they necessarily exercise the despised power of intellect,

* A recent example of this tendency—and of the dubious implications of Bergson's thought—is *Chart for Rough Weather,* in which Waldo Frank traces all the woes of the world to its religion of "empirical rationalism." He soundly criticizes the excesses of this faith, the neglect of the needs and the knowings of the organism as a whole; but when he proposes instead a religion based on the "organic real," the "true intuition," the "mystic truth," his logic collapses into logorrhea and organism becomes orgasm. Apart from the cloudiness of this "divine" knowledge, he does not tell us how to get at it, recognize it, distinguish it from the false or inorganic intuition, get it over to the unfortunates who do not have it. Worse, there appear to be various legitimate brands of mystic truth. All great religions and cultures, Frank says, have been premised on "myths that embodied the organic intuition of the race and of the whole man of the race"; but all are different, often contradictory, and how are we to choose among them? Briefly, although he is horrified by Hitler's version of the intuition and the myth, he leaves no intelligible standard for condemning it.

claim the rejected authority of reason. Alter the metaphor, speak of man reasoning or the reasonable man, and it is indispensable to any hope of civilization; to scoff at it is to clamor for race suicide. The natural end of human life, and now the necessary means of social life, is some rational organization of irrational impulse.

Perhaps the best way of getting into Bergson's thought, and then out of it, is accordingly through the thought of James. James was as hostile to all the schematisms and bureaucracies in thought, as devoted to the truth of immediacy, but he came closer to achieving a reconciliation because his primary aim was not so much unity as *totality.* He could admit the limitations of reason without handing philosophy over to intuition, he could admit that man has a good ear for monistic music and still keep tuning in on other bands. His scientific training held him closer to the "irreducible and stubborn facts," which if not strictly "immediate" are none the less given; he stayed with them instead of heading for the thing-in-itself. His functional view of truth was more consistently empirical and enabled him to keep discovering realities rather than wooing an ineffable Reality. In general, his pragmatism is more satisfactory than Bergson's intuitionism if only because it leaves the philosopher more work to do and more to work with.

My indebtedness to James, for vocabulary as well as idea, is apparent enough. But I should stress chiefly the general attitude —the temperament, if one wants—that he brought to bear on all problems. He will live, I believe, primarily as a humanist in the wide sense, not as a formal philosopher. Thus Peirce remarked that he was "phenomenally weak" in analytical definition or logical formulation—and as superior to most thinkers in his grasp of the practical bearings of an idea, its "cash-value" in experience. Thus as a psychologist he founded no school, worked out no systematic method, but his ideas reappear in the functional and dynamic schools, the introspectionist school, psychoanalysis, behaviorism, Gestalt psychology—virtually every significant development since 1900. In all events he was willing to put up with a "wild world," a world of appearance, change, variety, novelty, contingency, evil. He accordingly welcomed "any ideas upon which we can ride, so to speak: any idea that will carry us prosperously from any one part of our experience to any other part, linking things satisfactorily, working securely, simplifying, saving labor." Nevertheless he could also be frugal and firm; he achieved a rare union of tolerance and vigor, catholicity and conviction. "A bold riot of thinking," Perry summarizes, "qualified by a shrewd or inspired faculty for happy hits, and controlled by a scrupulous acceptance of experience's verdict—that is James' recipe." The humanist could have no better recipe.

Yet James had the defects of his virtues, and these are also instructive for the humanist. He was not always critical enough of the ideas on which he rode; he democratized philosophy and logic, but at some sacrifice of standards; he was somewhat too eager to make good in the American way, by free speculation and quick cashing in. The essential defect appears in his praise of contemporary philosophy: "It lacks logical rigor, but it has the tang of life." If most philosophies could do with more tang, no less pertinent is Morris Cohen's essay "In Dispraise of Life, Experience, and Reality"—as honorific instead of descriptive terms. Americans in particular could do with more logical rigor. Even in literature, where the tang is essential, we have had too much undisciplined thought and feeling, too much work in which form, clarity, precision, or penetration have been sacrificed in the interests of "real life." The philosopher, at any rate, cannot afford to cultivate Experience at the expense of reason or logic. It is an ambiguous term, which can include everything that happens or anything he wants; when he glorifies it, he is apt to blur the necessary distinctions between real and imaginary, knowable and unknowable, valid and invalid.

Hence, more specifically, James was too easygoing in his application of the pragmatic test. For thousands of years, Bertrand Russell comments, the Chinese have beat gongs to scare off the dog who causes eclipses by swallowing the moon; and the idea has apparently worked very well, by the pragmatic test it might be considered true. Causes and consequences are extremely difficult to determine, above all in the human affairs that especially concerned James. Moreover, he was apt to confuse logical with psychological necessity, the truth of a theory with the truth of a sentiment or belief. God, he said, means "that you can dismiss certain kinds of fear." So does the custom of beating gongs. But God also means that you can catch other kinds of fear, nobody can say what God has meant altogether; and whatever the consequences of a belief in Him, they logically prove nothing at all about His existence.

For James in turn, then, the best supplement seems to me the philosophy of Dewey. Dewey's "experimentalism" or "instrumentalism" is substantially the same as James's pragmatism. He fully recognizes the claims of immediate, concrete experience; he begins and ends with everyday life, the world of doing; he regards thought as a human, historical enterprise, and all ideas as "candidates for truth"; he opposes all principles or attitudes, whether derived from traditional philosophy or from traditional science, that tend to impoverish experience and restrict its possibilities. His thought (not to mention his style) scarcely has the tang of James's; it does not arise from so vivid and abundant experience, reflect so rich a personality. But he supplies the logical rigor that James lacked, and works out a complete, systematic philosophy. At all times he stresses the discipline of controlled inquiry, making the pragmatic test more like the experimental test of science and less dependent on happy hits. He pays more attention to the objective validity as well as the practical convenience of ideas, finds more room for truth as well as truthfulness.

No doubt Dewey himself is still not rigorous enough, especially by comparison with specialists in symbolic logic. Even the layman can detect some vagueness and looseness in his thought. As should be evident, however, what I have been calling scientific humanism is essentially the philosophy of Dewey (and is indebted to it to an extent, indeed, that only recently I began to realize). It should also be evident that I am in no position to criticize the rigor of his logic. So it is again merely to acknowledge the natural hazards of humanism that I make two broad reservations. His too easy assumption of the possibility of controlled experiments in social affairs is characteristic of his general attitude—a rather vague optimism, typically American, that leads him to minimize the difficulties of applying his method of intelligence. These difficulties are obvious enough in the humanistic studies; it is strictly as impossible to ascertain as to foresee the total consequences of any thought or deed. And also typically American is Dewey's emphasis upon ceaseless action, inquiry, experiment—the strenuous life. Allport has remarked that in his psychology he accounts much better for the shifts and developments in personality than he does for its stable structure. So in his design for living there is little place for simple sport or play, the contemplative pleasures, the easy look at life—the attitudes so admirably represented by Santayana. We can do more in and to the world that Dewey pictures, but we must forever keep doing. All is stress and tension; the passwords are "keep moving," "step lively," "watch your step."

Nevertheless these temperamental excesses do not menace Dewey's basic philosophy. They are excesses, in fact, of its soundest principle, his constant relation of values to intelligent action in the concrete context. Thus he points out that the most precious things in life are just the immediate, transitory, unstable, unique things; the general, uniform, and permanent that philosophers have considered the only reputable signs of Being are actually important to us only as the efficient means or conditions of these evanescent things. And thus he was able to make his profoundly significant criticism of the whole philosophical tradition. For over two thousand years the central concern has been with an essential, immutable, eternal, absolute reality; whether this was known by reason, intuition, or revelation is relatively unimportant. Dewey is one of the few philosophers to accept unflinchingly what philosophy has most obviously been—a cultural product (not to mention the product of a given temperament). It has not actually revealed the ultimate reality; it has interpreted the needs and interests of a given age. It has therefore made no real progress toward the goal it set for itself. Yet it represents progress in the same sense that civilization does: it has explored new possibilities in means and ends, expanded and enriched the significance of life. It could have contributed still more except for its obsession with a superreality. The gist of Dewey's philosophy, and the measure of his contribution, is the test he proposes for the value of any philosophy:

> Does it end in conclusions which, when they are referred back to ordinary life-experiences and their predicaments, render them more significant, more luminous to us, and make our dealings with them more fruitful? Or does it terminate in rendering the things of ordinary experience more opaque than they were before, and in depriving them of having in "reality" even the significance they had previously seemed to have? Does it yield the enrichment and increase of power of ordinary things which the results of physical science afford when applied in every-day affairs? Or does it become a mystery

that these ordinary things should be what they are, or indeed that they should be at all, while philosophic concepts are left to dwell in separation in some technical realm of their own? It is the fact that so many philosophies terminate in conclusions that make it necessary to disparage and condemn primary experience, leading those who hold them to measure the sublimity of their "realities" as philosophically defined by remoteness from the concerns of daily life, which leads cultivated common sense to look askance at philosophy.

Finally, the whole range of Dewey's thought—his logic, his metaphysics, his ethics, his esthetics, his essays on education and politics—is unified by a further intention that springs from his conception of the social responsibility of philosophy. It is the intention of achieving a harmonious integration in an age rent by discords and deep confusions. Though his primary instrument is the logic of scientific inquiry, his approach is not narrowly scientific, and though he is always critical of tradition, it is not merely modernistic. He retains almost everything in our heritage that the humanist would wish to preserve. He seldom lapses into the prevalent attitudes because of which, at the end, I part company with many who now call themselves scientific humanists.

These men are mostly scientists, and at worst they may talk like superplumbers. "Our expectation of living," wrote Lancelot Hogben, "has increased as we have learned to worry less about the good life and more about the good drain." Right now our expectation of living would be considerably greater if the world's leaders had worried a little more about the good life. When the scientist becomes less materialistic, he is still likely to take a pretty simple view of the social, humane uses of scientific knowledge. Thus Haldane once illustrated these uses by his decision to buy no more glassware, because statistics showed that the glass industry more than any other shortened the life expectancy of the worker; but there is some question whether his boycott affected the industry, and as much question whether glassworkers would appreciate a cut in pay or a loss of job in the short life left to them. Or when the scientist becomes farsighted, he is still likely to see what the situation logically calls for as what must happen—as it would in a laboratory. Arthur Compton declares that the growth of social coöperation is inevitable because the growth of technology has created a greater need of coöperation. By this argument, Kenneth Burke comments, 'It is hard to see why . . . the world should ever be lacking in virtue, for God knows we need it."

At best, the philosophers of science are still apt, like other philosophers, to soar to heights of thought from which the earth is hidden by clouds. In *Humanism and Science,* Cassius J. Keyser bathes everything in so rich a glow that nothing stands out clearly except his immense good will. Humanism, Science, Religion, Mathematics, Ethics—all are capitalized, all are reconciled by generous definition, all are then united in the "Dream of dreams," the vision of world unity; and Keyser gets more ecstatic at the thought that the true Ideal must of course be unattainable. In *The Promise of Scientific Humanism,* Oliver L. Reiser offers a far more acute analysis and a closer synthesis of modern scientific thought, in order to define an attainable ideal for a world very far from unity; but he concludes by prescribing "a new logic and a new language." He regrets that our illness is so grave as to call for "such heroic remedies." Yet even these do not seem likely to prove effectual; nor can we thrill easily to the strange scientific hymn with which he ends. "Throughout the entire realm of intellectual history," he writes, "only one idea can be found possessing the essential psychic motivation, the hereditary cultural prerequisites, and the scientific affiliations necessary to serve the race at its present juncture." This is the ancient idea of alchemy. Chemistry has given it a new lease on life by the discovery of radioactive phenomena; we now know that nature is actually producing a transmutation of elements. Reiser believes that in radiation or light is the secret of consciousness, "the real philosopher's stone and the elixir of life," and that this subtle essence is the means to a new world religion, the creation of a "World Sensorium." "That we should look to chemistry for cultural guidance is quite in keeping with the spirit of the times," he declares. I suppose it is. But chemistry will indeed have to work miracles of transformation before the race of man can be healed by a philosopher's stone and fired by hymns to a Sensorium.

All in all, science has enormously expanded our capacity for knowing, and has thereby affected our feeling and our striving. It has not made for a harmonious development of our powers, and alone cannot. We have still to assimilate all our knowledge, to feel appropriately and to strive wholeheartedly. Men of science properly boast that it is "deaf and blind to passion," and does not permit notions of value to obstruct the search for truth; yet passion is a fact and a necessity, the driving power behind all endeavor to create a better society, and no problem is more fundamental than the problem of value. Today in particular we need to remember that science is only a means to an end, for all too possibly it may prove the means to the literal end, of our civilization. Although it cannot be blamed for the abuse of its power, neither can it prevent the abuse. It does not itself give full directions for the best use of this power. And in such matters literary men are also worth listening to. They can help really to humanize scientific humanism.

2. THE RELATION TO LITERATURE

As we have seen, the philosophy of modern science makes possible friendlier, more fruitful relations between science and literature, on the basis of peace without victory. Unfortunately, this welcome news has not got around, or been generally welcomed. The old combatants still snap at one another. They still want victory.

Workaday scientists, to begin with, are seldom up on their new philosophy. They are habituated to the hard fact, in their laboratories they do not feel that they are dealing with fictions; and so they are prone to identify the poetical with the impractical, the imaginative with the imaginary, the fictional with the false. Similarly thinkers who aspire to be scientific are prone to dwell on the primitive origins and psychological poor relations of art, regarding it at best as a harmless narcotic, possibly useful for community hygiene, or like sport as an outlet for surplus energy. Seldom do they admit it as a legitimate, mature, important way of knowing and dealing with reality. Literary men, on the other hand, are apt to be even more supercilious toward the upstart science, and still assert that its truth is a relatively poor kind of thing. The knowledge in art is superior, says Herbert Read, "because whilst nothing has proved so impermanent and provisional as that which we are pleased to call scientific fact and the philosophy built on it, art, on the contrary, is everywhere, in its highest manifestations, universal and eternal." Offhand, Archimedes' principle of the lever seems as universal and eternal as anything in Greek art, and as for the truths that are more permanent and unprovisional than a scientific fact, we might ask

Read to name one; but he could no doubt escape on his "highest manifestations" into some spiritual realm where such questions are impertinent. Or he might refer us to the whimsical idea expressed by Robert Frost (in a rather low manifestation):

> At least don't use your mind too hard
> But trust my instinct—I'm a bard.

Now men of letters have a legitimate grievance today. Despite the growing distrust and fear of science, it still dominates the world of thought and puts all other interests on the defensive. Its claims to authority are more generally admitted than are the claims of literature; it does not suffer from the competition of literature as literature has suffered from its competition. For just this reason, however, men of letters are foolish simply to oppose their truth or their instinct to the unquestionably efficient use of mind in science. If only as a matter of strategy, they would do better to establish a clear connection. The familiar distinction between the logical, factual truth of science and the imaginative, spiritual truth of art is agreeable enough—until someone asks, "What of it?" The problem is to give these symbols more substantial, negotiable meaning, to bring these species of truth into more usable, convertible relation. It is finally to make both the scientist and the man of letters more social and sociable. And so we might begin by pointing once more to their underlying kinship.

Forget the abstractions "science" and "art" and all the high talk about them, watch the individual scientist and artist at work, and they look more alike than either is like any other worker. The artist too is a ceaseless experimenter, and his artistry is as rigorous a discipline as a scientific inquiry. The whole process of creation is an inquiry, a constant exploration and trial of possibilities. Nor are utterly different faculties involved in this process. The scientific mind is not a mere logic-grinder, turning out truths with remorseless precision. It too feels its way, has flashes of insight, leaps to conclusions—arrives intuitively at an intuitive goal. Although the final experimental test is all important, what is tested is a hunch, an inspiration, a dream. All the great scientific theories are great imaginative feats.

Likewise science and art necessarily begin with the same immediate data. The scientist may try to eliminate "secondary" properties, but a sensory perception is always his final referent and the means of his experimental test; the poet may talk of timeless essences and pure spirit, but he always appeals to and through the senses. They both use sense to make sense. There is still some resemblance, moreover, in their finished products, which are statements or expressions of relations. Scientists like the formula H_2O, not because it tells what water really and truly is, but because it enables them to relate water to solids, gases, and the whole physical world. No more do poets distill the quintessence of water; they simply relate it to the whole world of felt experience through its sensuous qualities and emotional associations. Metaphor has something like the economy and efficiency of the formula H_2O, drama works out something like an equation, artistry in general is an effort to seize significant relations.* Altogether, the work of art and the scientific theory alike give us significant form: an ordered pattern, a composition of experience. The value of truth is also imaginative, the value of beauty is also practical.

This is not to say that science and literature come to exactly the same thing. Most significant, naturally, are the differences that give each its peculiar usefulness and make it impossible for one to perform the whole function of the other. The point is again that these differences are not antitheses, and properly do not call for elections or referees. But it is also this most important implication, that art too is a real way of knowing the real world. With the concept of a soul independent of matter, esthetic experience may indeed be etherealized, purified clean out of natural existence. With the concept of an organism inseparable from its environment, the artist has a natural way of getting in touch with actuality. Since the meanings of experience are as wide as consciousness itself, it would no doubt be helpful to restrict the terms "knowledge" and "truth" to those meanings alone that can be exactly formulated and strictly verified; philosophers, theologians, scientists, and poets might then go about their business, feeling as superior as they pleased but at least not demonstrating their superiority by calling different things by the same name. Unhappily, however, these are honorific terms. The distinction naturally suggests that the meanings in literature are *untrue*. Those who value

* Modern poets in particular, as I have noted elsewhere, have adopted principles comparable to those of modern scientists. The main trend has been toward "dynamic composition" or "organic structure," a full realization of the value of the "pattern," in which metaphor and imagery are not merely decorative but "functional" or "integral."

these meanings are then driven to claim for them some more essential kind of truth. And so criticism has been torn between these extremes, that art is a superior knowing or that it is no knowing at all.

Newman distinguished two modes of understanding, "abstractly," or through known truth, and "really," or through experienced truth. The "real" mode, of course, was the better one. Ramon Fernandez now tries to make the distinction uninvidious. Art, he says, "does not make reality known, it makes reality exist"—it is "a realization of the real"; but he goes on to emphasize the necessity of reason for creating the reality to be known and checking up on its realization. From here, however, it is an easy step to the famous dictum of I. A. Richards, that poetry "tells us, or should tell us, nothing"; it merely induces fitting attitudes. Then Max Eastman finishes the job: if the poet can tell us nothing, he cannot be trusted to induce such attitudes; his function is merely to vivify and prettify consciousness. By this time we are out of sight of the actual intention of nearly all poets, the actual experience of nearly all readers, and we are left no clear grounds for resisting the conclusion that the greatest writers have only made a pretentious fraud out of a childish make-believe. We have arrived, quite logically, in plain absurdity. And the only logical escape is to go all the way back to our premises. It lies in the assumption justified by our experience, that poetry, in a matter-of-fact and no mystical sense, by natural and not transcendental means, does make good sense, does make reality known, does tell us something.

The most familiar approach to the understandings in literature lies through "intuition" and "imagination." Nobody denies the artist these faculties, and they can get him at least within hailing distance of knowledge. Taking intuition in its broadest sense, we may say with Croce that conception is empty without it just as it is dumb without conception. Similarly imagination, conceived as the synthetic power of the mind as a whole, brings literature into relation with all ambitious thought and significant enterprise. Call the imaginative creation by its worst name, "wish fulfillment," and it is still a force in knowing and doing; for the wish is often reasonable, is always the father of the deed as well as the thought and the dream. But the trouble is that these are pretty slippery words. "Imagination" is difficult to grasp except by its own metaphors, apt to lead chiefly to further imaginings; "intuition" has become still more elusive since Bergson, and in his sense cannot be defined or discussed at all without being falsified. At any rate, like "reason," "emotion," "will," or what have you, both are involved in the mental processes of scientist and artist alike.

Hence it is more profitable to distinguish instead the different objects of knowledge in science and literature—not how but *what* they tell us. Science, once more, never tells all, never tries to tell everything about anything. It has found out so much because it has so rigorously defined and limited its objects. And the primary subject matter of literature is precisely all that science leaves out: the individual, the particular, the concrete—the "world's body." Although writers make use of abstract knowledge, and are all too likely to strain for "higher" truths, actually they differ from philosophers and scientists in that they do not so consistently go higher, but stick to immediate experience; as Ransom says, they are "prodigious materialists." Hence they still have the whole substantial world as their province after the scientist has said all he wants to. The romantic notion of the poet as one dreaming on things unknown, the shapes of things to come, points to a source of genuine interest; but it also leads to the common misconception of poetry in full retreat as science advances into the unknown and shapes the things to come—of all literature living on sufferance, feeding on leftovers. Actually, science leaves off as well at the most familiar, the immediately *known*. It is interested in how and why things happen, not in our direct experience of them. Literature is more interested in the happening itself, the experience for its own sake.

Specifically, the artist reports the qualities of experience. They are the qualities of a physical world that is extraordinarily varied in its appearances, extraordinarily uneconomical of possibilities of sensation. They are found as well in situations, personalities, life histories, civilizations—they are the character or configuration of any organic whole. They are involved in such unmathematical truths as that $1 plus $1,000 is a lot more money than $1,000 plus $1. But in all events the artist's report is not merely fanciful or necessarily vague. To call a color bright or pale, a landscape smiling or austere, a temperament cold or mild, a philosophy harsh or humane, is to make an intelligible reference to a significant fact of experience. In general, artists are our specialists in deciphering, discriminating, and communicating all such impressions, which are also guides to rational thought and behavior, important not only in the art of living but in the practical business of life.

We may accordingly qualify even the truism that the language of science is more exact than literary language. It is exact in that it is unambiguous. A proper scientific phrase makes a single definite reference, it can be replaced by an absolute equivalent, it can be transferred absolutely intact into another language; its complete meaning has been completely agreed upon in advance. A poetic phrase (as distinguished from the more or less scientific phrases that have their place in poetry) cannot have an absolutely identical meaning for all readers. The poet tries to convey sensations that are never pure or simple, personal impressions that are strictly unique, and he deliberately invokes feeling tone, connotation, sound effects—all the orchestral devices that enrich meaning but also blur its edges. By the very measure of his success, his phrase cannot be exactly defined, replaced, or translated. Nevertheless it is the most efficient kind of phrase for the purpose of communicating these subtle, complex impressions. In a way the poet achieves an even finer precision because the phrase is unique, expresses what has never been said before and can be said in no other way. But in any event exactness is no guarantee of truth. The strictly exact scientific statement may be strictly false—scientists are forever testing and scrapping such statements. The strictly inexact poetic statement may be broadly true—the more likely as the poet does not strive for scientific truth.*

Briefly, it is no mere accident or carelessness of speech that the word "symbol" denotes at once the mathematician's *x* and the poet's figure, a logical sign and a flag; for these are related means of communicating and controlling experience. And here we approach the unfashionable, often embarrassing, but fundamental idea of literature as a criticism of life. Like it or not, literature has in fact always been a "power of conduct." It has schooled pur-

* One moral to be drawn from this distinction is the necessity of poetic language for some purposes of literary criticism. The semantic ideal is appropriate when the critic is stating the intellectual content of a given work, abstract ideas about it, general principles of criticism. It is not appropriate when he is trying to convey the quality of a work, its esthetic meanings and values; here the ideal is the precision of poetry, not of science. No doubt critics can usually be trusted to be lyrical or ambiguous enough; but there is some point in urging this moral in these positivistic days, when too much prose is merely businesslike, and criticism often cultivates a factitious exactness by the use of technical jargon or the evasion of esthetic issues.

pose and desire, inculcated values and ideals, which is to say the ideas that men can sing about. It has helped to define and to realize the meanings of whole ages—the glory that was Greece and the grandeur that was Rome. One may say that the writer should not deliberately aim to instruct or edify; literature should not begin as a criticism of life. But it cannot help ending so.

Perhaps the safest way of getting at this final consequence, without feeding the passion for essential or elevated truth, is to say that the writer, even when most impersonal, always expresses or implies an attitude toward his imaginative object. The scientist only points to his facts; officially he expresses no other attitude than the hope of deriving more facts from them. There is not even an imperative mood in the grammar of science, but merely a conditional tense—*if* this, *then* that. In daily life, however, we constantly take up attitudes, translate indicatives into imperatives. We point with pride, with alarm, with disgust; we deal with facts in terms of "Hurray!" "Ouch!" "Phooey!" The main genres of literature are so many extensions and refinements of such terms. Comedy, tragedy, epic, lyric, satire, burlesque—all suggest ways of looking at and dealing with a situation. By these ways they satisfy the poorly defined, much abused, but very real "emotional needs" of man. And it is accordingly important that the writer, *as artist,* is equipped for this business. Through his sensitiveness to the qualities of experience and to the concrete situation, his concern for the individual, his command of the necessities of the inner world, he can indeed induce fitting attitudes. Often, of course, his criticism of life is narrow, confused, or distorted, and especially as it becomes explicit it needs in turn to be criticized; but his fallibility no more invalidates his mode of thought than the constant junking of scientific theories invalidates scientific method.

As far as it goes, scientific knowledge is unquestionably a surer guide to appropriate attitudes. Just as unquestionably, it does not go far enough. The uses of language are again the key to its limits. Scientists have created a highly efficient language by emptying their symbols of all irrelevant associations, more especially of all reference to attitude and emotion; they aim at a complete neutrality for the sake of objective clarity. Poets load their symbols with as many meanings as they care to or can get away with, more especially with references to attitude and emotion; they also aim at objectivity, but they achieve it by including and finally clearing up rather than by cutting out in the first place. Their language is more live, has to do much more work to earn its living. In the words of Kenneth Burke, the scientific vocabulary *avoids* drama, the poetic vocabulary *goes through* drama. The one never gives emotion a fighting chance, the other takes on all the emotion around in order to control it. For both our knowings and our doings, then, the language of science is insufficient; it cannot express all the meanings found in experience, and it cannot handle all the problems raised in experience. For outside the laboratory it is impossible to be semantically pure and chaste. We cannot simply eliminate attitudes, emotions, values, desires, multiple and messy meanings. We have to size up the whole situation as it is, take on all its meanings, go through it—more completely learn and *earn* by experience.

In this perspective, it is the realistic, unflinching scientist who simplifies the issues, dodges the difficulties, at best runs away to fight again another day. It is the wishful artist—the dreamer, the neurotic, the escapist—who uncompromisingly faces the facts and accepts all their consequences. Today the situation calls for such an upside-down look; we have first to deal with an actual disharmony and imbalance. Yet the object in taking such a look is finally to get clear of these invidious descriptions. We cannot hope to right matters until we realize that properly science and literature are complementary, not competitive activities. The triumphs of one never need to be at the expense of the other.

To repeat, the actual influence of science on literature so far has in many ways been unfortunate. A hatred or fear of its works moves some writers to needless negations or retreats; an awe of its works keeps others on the lower levels of realism, where truth is literal and fact unvarnished. When they outgrow these simplicities they still have to contend with masses of undigested knowledge and unnaturalized attitudes, and with the ruling habit of scientific abstraction. It is very hard, in this mental climate, to keep the eye steadily on the esthetic object, to remember that art is less an abstraction from concrete experience than a condensation of it, to distinguish always the positivistic from the poetic linguistic ideal—never to confuse the materials, logic, and language of science with the materials, logic, and language of art.

At the same time, science has also contributed much to literature, given writers new insights into their old story of the ways of human nature. It can contribute much more, once it has become assimilated in the living tissue of thought, and men realize that the natural world it pictures is less alien to the human spirit than the world pictured by many poets and theologians of the past. Apart from all the specific implications discussed in this work, it has enriched the whole background of thought and feeling. For however dangerous abstract knowledge may be to specialists in it, its final effect is to deepen, expand, refine, and fertilize our direct experience. Science can make possible a greater literature for the same reason that a child or a savage is incapable of rich esthetic experience. It is the reason implied in Elie Faure's dithyramb on civilization: "The sum total of the spirit, the sum total of the sensible universe, become thus more closely linked, more complex, the sources of emotion, of effort, of drama, are opened in a hundred still unexplored corners of the brain and even of the heart." After such visions, it is a little disheartening to contemplate the actual modern work of art; but in Thomas Mann, at least, one can contemplate the possibilities.

Finally, however, the chief value of science is of course as a co-worker, not a servant of literature. The knowings of art are real, valuable, necessary; but they are not utterly reliable or sufficient. In celebrating the recorded values in poetry, its resources of noble idealism and practical wisdom, critics are likely to forget that its profoundest insights are accompanied by very dubious assertions, its sublime imaginings by absurdities, its prophetic visions by many more wrong guesses—that a composite of the world's greatest poetry would make a pretty unintelligible picture of reality or bible for dealing with reality. Poetry as poetry offers no clear criterion for making these all-important distinctions. Analytical reason is necessary to make its "imaginative truths" thoroughly intelligible and to determine their pragmatic validity, which is to say the probable consequences of action based on belief in them. It makes clear the meaning and the imaginative value of the witches in *Macbeth,* for instance, but it can also warn us not to behave as if there were witches. In short, literary men who oppose the "merely" practical values of science to spiritual values have essentially the same attitude as ladies who just love culture. All savage societies had spiritual values; but there has been and can be no civilization without at least the rudiments of science.

For thinkers at large, however, there remains more need of stressing the other end of the give and take. No doubt the direct contributions of literature to science are often exaggerated. If some scientific theories, such as Freud's, are systematic formulations of ideas long familiar in literature, it is too easy to discover such imaginative anticipations, given the poet's old custom of speaking like the Delphic oracle. The indebtedness of scientists to literature is more indirect. Yet it is greater than they usually appreciate. It might well become still greater, especially as their triumphs induce them to be oracular too.

Aside from its general contribution in also widening and enriching the whole context of scientific thought, literature can offer a pertinent criticism of this thought. One need not believe that the poet sees with larger, other eyes to respect his intuitions. The nineteenth-century poets who felt that something essential had been left out of the mechanistic account of the world were often cloudy and immoderate in their criticism; but it is now generally agreed that they were right. For scientists are apt to become the victims of their habit of abstraction, and their ascetic way of life. Precisely like their traditional enemies, the philosophers and the theologians, they tend to find "illusion" and "error" where in daily experience we most naturally, positively find reality and truth. Because their system of knowledge is governed by the demands of their logic, not by the claims of our experience or our practical needs, they tend to discredit our simple understandings. And for their own purposes they need to be reminded that their account of the world is necessarily partial, so that they may more freely alter or expand it to find room for new facts and new possibilities. Especially in the sciences of man, the chief reason why their concepts have so often been inadequate, at times even a little absurd, has not been faulty logic or incomplete knowledge, but their refusal to admit relevant data that writers take for granted.

Through literature, moreover, scientists can get a perspective on the natural, social function of their activity. In their own tight little community they may forget or even scorn such homely considerations. As they rejoice in their triumphs, they may fail to ask themselves the pertinent question: triumphs over *what* and to what *end?* Even as they piously preserve the names and works of the great scientists of the past, they may forget that this is a poetic act, strictly superfluous—the great man's contribution to knowledge is already embalmed in their formulas and by definition can have no personal quality. Poetry, "the impassioned expression of science," has indeed paid the most eloquent tribute to the faith on which science rests, the faith in the *value* of truth seeking. But the poet is also likely to perceive the limits, bearings, and consequences of scientific truth. Without this perception its value cannot be fully realized, and the faith may become one of the most arrogant, despotic of religions.

In any event, literature is a necessary supplement. "A blind man can know the whole of physics," Bertrand Russell has said; but we plainly are not content to live as blind men. We want the appearances, to us the world also is as it appears. We are literally growing more conscious, Oliver Reiser declares: we are more thoroughly aware of our environment, "we react to a wider range of stimuli in space and time," and so we may experience an "increased intensity of inner life." If so, we must appeal to the artist. The scientist as scientist is no such sensitive plant, and experiences less intensity than did the prophets of ancient Israel. Often, indeed, he views the inner world as a kind of lunatic asylum of feelings and fancies; he still tends to make a sharp distinction between inside the skin and outside the skin, as if this flimsy envelope were an impenetrable wall. His passion for intellectual purity, a virtue free from all taint of the symbolically unclean, in effect becomes a theological scorn of the flesh. It is therefore the more unfortunate that poets have often been held up to this same ideal, urged to avoid the "sensuous" and the "worldly." Nevertheless they have never been utterly corrupted by such idealism. Their natural impulse is a loyalty to concrete experience, to the living consciousness—to the sensory image, which can express the most highly spiritualized ideas just as it is the basis of the most remote scientific abstractions.

In general, the area of significant meanings must always be far wider than the area of scientific truths. So must the beliefs we live by far outrun the scientific evidence. "There is at bottom very little difference between men," said the old carpenter whom James was fond of quoting, "but what little there is is *very important*." Few would deny this importance. Yet the statement is not a scientific truth, it cannot be handled by symbolic logic. It is the kind of belief asserted or embodied in literature; and it leads to all the ancient platitudes that express the importance of literature. We still need to *accept* the universe, as known and as unknowable. We still need imagination, sensitiveness, sympathy, good will, the "heart to do it." The prebeliefs and the overbeliefs required by the nature as well as the state of human knowledge, all the faiths and ideals that cannot be put to a strict operational test, become more important as science places more and more power at man's disposal.

In short, the very triumphs of science only accentuate the further need of literature, bring us back to the fundamental complementary relation between them. With its immense intricacy, modern society is obviously dependent upon the efficient system of communications that science provides. With its immense heterogeneity, it also has especial need of an art of communication, a full and moving expression of the common meanings that hold a society together; and literature can not only realize the real but socialize it, effect not only communication but communion. For science, the Jehovah of the modern world, has not created man in its own image; it cannot itself make him at home in the world it keeps remaking, with no day of rest. Many thinkers are able to find their way around, and have become oriented in conceptual terms. Many fewer have brought this knowledge back to life, got it into their habits of seeing, feeling, and doing; with most it has gone to their heads and stayed there. And here the extremely difficult problems forced on literature are also the great opportunity of literature. It can assimilate the meanings that science merely formulates, reclaim them in terms of feeling and sentiment, fuse them with older and deeper meanings, naturalize and humanize them. It can lead from a logical consistency to a vital coherence, from a logical synthesis to a vital harmony. It can help men really to own the world created by science.

This is a mighty task, and certainly not the solemn responsibility of every writer; the simple meanings are still pertinent, the old themes still valid, the old appearances still real, and literature must take account of all of them. Nevertheless writers perforce do tackle the new job, however unconsciously, indirectly, tentatively, as they try to make sense of the world in which they live. Thomas Mann's work is again a striking example of the problem, the opportunity, and the possible achievement. In *The Magic Mountain* he "took stock" of the whole intellectual situation, from a mountain top. In the *Joseph* story he is approaching, through the eternal rhythms and patterns of man's being, in a spirit at once profoundly ironic and profoundly reverent, the realization of the complete union: of culture and nature, past and present, masses and man, poetry and criticism, science and ritual, sanity and sanctity.

Ultimately, the whole issue of literature and science comes down to the age-old problem of all thought and intelligent living. In Santayana's words, this is "to unite a trustworthy conception of the conditions under which man lives with an adequate conception of his interests." Art deals more directly with the subjective interests, science more directly with the objective conditions. But neither deals with purely private or purely public events, neither gives the soul-in-itself or the thing-in-itself. Although science seems obviously closer to the facts, its whole logic and language constitute an elaborate *As If.* Although literature seems obviously to be remolding the world to the heart's desire, it more closely represents the immediate *As Is.* Imaginative literature is also matter of fact, practical science is also matter of fiction. Both seek freedom, through control of the necessities of the inner and outer worlds, and both must first obey in order to command. Both are on the "growing edge" of things, united in a common endeavor, which is to naturalize man upon this earth.

No doubt there will always be a deal of misunderstanding and friction between men of science and men of letters. They are alike open to pretensions of self-sufficiency, religions of art-for-art's sake and science-for-science's sake, simply because any consuming activity is its own justification, and thoughts of social function are afterthoughts. In the endless competition between major interests there is never a perfect balance. But it is therefore the more important to remember that between major interests there is never an absolute distinction nor an absolute antagonism.

3. THE RELATION TO RELIGION

Most of the men who are deeply troubled by the decay of religious faith nevertheless seem strangely absent-minded. They assert that religion is the crying need of the modern world—but they seldom tell us how to get religion. They seldom come to grips with the obvious problem, which arises less because contemporaries are hostile to belief than because they are simply indifferent or simply unable to believe. They seldom tell us precisely what we should believe, and on what grounds. When

T. S. Eliot prescribes religion, he means the Church, and by the Church the Anglican Church; if his solution is rather arbitrary, at least we know what he is talking about. When most others argue for the necessity of religion, we do not know. Even the terms of the necessity are unclear. It appears to be now for intellectuals, as a sanction for their values, and now for the masses, as insurance of their good behavior. It appears to be now logical and now psychological or moral. Or it may be no strict necessity at all but only something very desirable, like the classless society. Altogether, the many earnest voices call to mind what someone said of Coleridge: they make a wilderness where they cry.

As usual, the immediate source and sign of confusion is linguistic, a matter of definitions. Religion has unquestionably played a vital role throughout history. It has satisfied deep needs, and we may safely assume that these needs themselves have not vanished. It has been identified with emotions, attitudes, interests, values that still appear essential to well-being. These truths, however, are self-evident only when unspecified. What we have to deal with is not Religion but a multiplicity of religions, and in this multiplicity the common element, the necessary condition, is by no means self-evident. We cannot easily assume, for example, that religion requires a belief in a supernatural power or being; some recognized religions have not invoked such a power. Nevertheless most men, at least in the Western world, do assume just this. To intellectuals as to simple citizens, religion usually means a belief in God. So here is the immediate issue.

Now contemporaries are apt to dismiss the supernatural hypothesis too abruptly or contemptuously, again by the simple expedient of calling it "meaningless." Their realism lands them in the sheer folly of assuming that we should accept only what is capable of absolute proof—an assumption that would automatically destroy almost all the values that all men live by. First and last principles are beliefs, not fixed truths, and as James observed, there is "no scientific or other method by which men can steer safely between the opposite dangers of believing too little or . . . too much." At the very least, God remains the most immense and splendid of all metaphors. Samuel Butler spoke more wisely and more reverently than he knew when he said that an honest God is the noblest work of man.

Yet it must also be insisted that the supernatural is *not* capable of absolute proof. The too familiar arguments still crop up. Men hold that because we are impelled to think of something, even because we *can* think of something, it must exist; and so they can prove whatever they have the heart to, and the moon really is made of green cheese. Others triumphantly misinterpret the principle of indeterminacy in physics, under the ancient illusion that if certainty is driven out of all natural knowledge it must settle in theology. And mystics continue to believe that because they seem to be out of their heads they must be in touch with the Godhead, instead of back in their senses, or that an ineffable experience is necessarily an experience of the divine, whereas nothing is strictly more ineffable than, say, the taste of spinach. In short, God must literally be taken on faith. The actual problem, again, is that more and more men are unable to hold this faith. The objection to most arguments for God is that they fail to meet the given terms of this problem.

At once the weakest and the most cogent arguments are accordingly those, not for religion in general, but for *a* religion, for the Church. They are cogent because the Church is a fact, a solid institution, an ancient rock—because it is *there,* waiting to serve the needs of men as it has done for centuries. This is the argument, for instance, of Michael Roberts in *The Modern Mind.* He affirms that men must have some absolute authority, to relieve them of the intolerable responsibility of making all their own vital choices, and he recommends the Church as the oldest, surest authority. Plainly, many contemporaries are unhappy in their emancipation, do find its responsibilities intolerable, and would be glad to stand on a rock again. But Roberts does not take care of these unfortunates. He does not relieve them of the responsibility of which Church to choose, not to mention their misgivings about the accident of birth that makes Christianity seem a more reasonable choice than Buddhism or Mohammedanism. Above all, he does not demonstrate the validity of this authority. Of the criteria of truth—"correspondence, coherence, and vitality"—he asserts that religious beliefs satisfy the first and third; but he merely asserts. Offhand, Christian doctrine seems coherent enough. The trouble remains that for many men it does not correspond to reality and has lost its power of vitality.

Faith, said Locke, is "assent to a proposition . . . on the credit of its proposer"; and the credit of organized religion is no longer good. Although the mass of practices, beliefs, codes that compose Christian dogma has a solid core of tested wisdom, it is naturally overlaid with crusted rationalization, streaked with irrelevant or discredited ideas. Even in a less critical age than this the dogma would raise difficulties, for it embalms the ignorance, prejudice, and superstition as well as the wisdom of bygone centuries. These difficulties, moreover, are not only intellectual but ethical. In one aspect the whole official business of worship and service of God—the petitions, flatteries, compliances, covenants, indulgences—is frequently condemned as a sordid contractual machinery by which men bargain for special favors, gratify their somewhat monstrous obsession with an immortal individual existence. Certain doctrines—such as infant damnation—now seem simply revolting. And at that the dogma is more attractive than the historical record of the Church that is to make our choices for us: a Church that has justified serfdom and slavery, set a long record of bigotry and oppression, bred hatred as well as love, and far more often opposed than led not only intellectual enlightenment but social reform. Today it still stubbornly resists efforts to liberalize its precept and humanize its practice—as to preserve its traditional authority it must resist. This is by no means, of course, the whole story. The Church has also produced many saints, inspired much noble effort, symbolized the loftiest ideals of men; and like all living things it *has* grown and changed. In a humanistic view it is a worthy symbol of the "tragic dignity of history," the strange, terrifying, wonderful capacity of man for good and evil, and for good-in-evil. But its disciples insist that humanistic standards are inadequate, when not simply presumptuous; and we may then fairly complain that the Church is an all too worldly, human institution to provide absolute sanctions. We may demand of the disciples the elementary responsibility of considering religion as it is found, not merely in the abstract or the ideal.*

* In *True Humanism*, Jacques Maritain attempts to escape this dilemma by a distinction between the historic Church and the "authentic Church" behind it. All that is dark and deadly in the theological record was the work of the historic Church; but the authentic Church was never involved in these shocking abuses, nor is its sacred authority a whit impaired by them. Unhappily, however, he provides no clear objective criterion for telling these Churches apart. The individual Catholic appears to be left with the awful responsibility of deciding which one has authorized a given policy—a responsibility that I believe is not provided for in his creed, and that historically, had he protested too strongly against now admitted abuses, would have laid him open

Most of the literary apologists for religion, at least, have little faith in it as it is found. They wish that the Church (whether Catholic or Protestant) had more spiritual authority; they rarely urge that it be given more worldly power. They want to keep religion out of government, education, or any position where it could effectively influence social life. Similarly they try to keep it on the lowest possible theological budget. Wishing to preserve the poetry and wisdom of Christianity, they play down all the dogma. It is an amiable, a pious effort. But it betrays them into a fundamentally false position. They are trying to smuggle in God without paying the intellectual import duties, they are becoming dilettantes in Christianity—they are themselves proof of the actual decay of faith. T. E. Hulme's attitude is sounder for the same reason that it is a little startling. "It is not . . . that I put up with the dogma for the sake of sentiment," he wrote, "but that I may possibly swallow the sentiment for the sake of the dogma." Although literary men are attracted mostly by the sentiment, what they want at heart is the dogma, the ancient rock; and some unqualified belief is in any event indispensable to a vital religion. Yet any dogma is still likely to stick in their throats. (Even Hulme's statement suggests more a desire to believe than a positive belief.) They wish to predicate nothing about God and drop the whole matter once they have assumed His existence; they are uneasy when official Christianity, in its realism, insists upon the absolute truth of its doctrines; they cannot bring themselves to say flatly that Christianity is *true*.

Here is also the difficulty with the pragmatic religion proposed by William James. Our faiths must be "hypotheses," he declared, and must never "put on rationalistic and authoritative pretensions." Such faiths, I should say, are unobjectionable. But they do not answer what is reputedly our crying need, they are very different from the world's great religions; none of these have offered their main principles as speculative or uncertain. Similarly with the view of religion as a sublime fiction. John Crowe Ransom argues that its metaphors and myths should be

---

to excommunication by the only Church he could be sure of. And though Maritain is an exceptionally humane, liberal apologist for Catholicism, often speaking out of the humanistic tradition, he still represents attitudes that through history men have learned to distrust. He admits religious freedom, for example—but only as a necessary evil in a still heretic world; he merely tolerates tolerance, he invariably criticizes all criticism in matters of faith and morals.

taken, not as facts, but as symbols of an adequate attitude toward life in all its possibilities. Here is indeed the source of their enduring value.* Yet a purely symbolic religion is on the same footing with science and art, and cannot offer the security of the old-time religion. Meanwhile Ransom's view also implies the objection to institutionalized religion. Its myths are always hard to alter or discard, because men naturally take them for facts. It often defeats the purpose of myth, which is to deliver men from despotic fears and obsessions, for it delivers them into new fears and obsessions. The fictions of science are more reliable for their given purpose, the fictions of art are freer and more plastic; religion has generally tended to attack the one, to narrow and rigidify the other.

This brings us to the positive dangers of a belief in the supernatural. It often leads men to stigmatize their natural interests as "worldly" or merely practical. Their "spiritual" interests, the "finer things of life" that even the Calibans aspire to, then come to seem unworldly and impractical. Hence religion has also helped to intensify the materialism it deplores: the obviously practical interests are unillumined by other values and become in fact gross, the obviously spiritual interests are so pure that they become strictly immaterial or ineffectual. In general, the traditional religious attitude tends at once to debase man too much and to exalt him too much. Men worship their Creator by anathematizing His creation, cursing all the sons of Adam with original sin, cursing their natural home as a devil's snare or a vale of tears. Nevertheless they insist on the radical imperfections of man and nature only to assure an eternal order, in which they hold a high position; they belittle their powers for the sake of the greater glory of God, but theirs is finally the glory as His favorite wards. This blend of servility and arrogance is the more striking when churchmen condemn the sinful pride of naturalists, whose attitude would appear to be more modest. It is at any rate fair to remark that the religious attitude too often obscures the real problems before us, or even encourages a shirking of these problems. If the doctrine of Original Sin may inspire a needed humility, it may also be a way of avoiding personal dis-

* Orientals, it should be added, find it easier to take God as a reflection of the human soul, not an absolute objective reality. So do some Western mystics. "I know that without me God cannot live a moment," cried Angelus Silesius; "if I am destroyed, He must give up the ghost."

comfort and responsibility, a too facile, self-righteous reconciliation with folly and evil. In social affairs, many of the godless or the god-forsaken are both more charitable and more resolute than T. S. Eliot.

Yet it remains no less true that religion has been a symbol and a source of very valuable attitudes. The attitudes themselves are of even more vital concern now that their traditional authority has been weakened. And here, as I see it, is at once the pious and the realistic approach to the problem. It is first to get at the essence of religious experience, the actual common denominator of the world's great religions. It is then to consider whether this experience cannot, after all, be had on modern terms.

Now the remarkable diversity of doctrine that is at first bewildering, or even a little shocking, is also comforting. It does not bother the self-righteous, and it makes plain to everybody else that the religious cannot be identified with any specific dogma. It suggests that the religious is to be looked for in a quality of experience or state of mind rather than a belief, that the belief is at bottom more the result than the cause of the experience—that what we are after here is more psychological than logical. The very general belief common to the world's religions is that there exists some larger, higher, unseen power that deserves reverence. But the root experience, the experience that makes religious belief precious to sensitive men, appears to be a sense of oneness with an immense whole: a feeling of complete harmony that at crises may lead to a lasting conversion, a coming to final terms with life, and at its most intense is the mystical experience, but that always works to steady and strengthen men's purposes, deepen their sense of values, support their ideal aspirations, enrich the whole significance of their lives. Altogether, the only clearly universal, necessary condition of religion has been adequately defined, I believe, by William James as the "wider self through which saving experiences come."

Usually the unseen power has been conceived as a supreme being, a deity outside of nature. Usually—but not always. There have been innumerable conceptions of its nature, and of the behavior it demands or deserves. "All that the facts require," James went on, "is that the power should be both other and larger than our conscious selves." He also believed that naturalism can offer "no clear explanation" of these facts of religious experience. And here I take issue. The naturalistic explanation is far from complete or certain; yet it is satisfactory enough to make the supernatural hypothesis unnecessary, at least as clear as any supernatural explanation, certainly less mystifying than such accepted doctrines as the Trinity—not to mention the whole stupefying theological word game that somehow got started from the assumption of the total infirmity of the human mind and the absolute incomprehensibility of the Divine Mind.

James himself suggested that the "wider self" is the unconscious self. The saving experience seems supernatural because it cannot be summoned at will and runs its own course when it does come. But the unseen power lies in the depths of the human mind, the feeling of oneness is with the whole history of man on earth. Religious experience may be conceived in terms of man's natural impulse to form and rhythm, as a heightened consciousness, an immediate realization—imaginative, intuitive, organic—of the reality now conceived by science. It is a sense of a cosmic Gestalt, of an immense continuum in space-time. In simpler terms, man could be expected, as a child of nature, to have some feeling for his parent. We do not need the assumption of a supernatural power to account for human limitations, forces beyond human control. All our knowledge and experience lead us to expect these as the given conditions of life in this world; and if the phrase "this world" is the epitome of all mystery and all sorrow, both are deepened by the assumption that an omnipotent Being chose to create such a one. Neither do our values require a supernatural explanation. To believe that the natural order is determined by our moral order is presumptuous; it is to conceive, Santayana remarks, a world where parents are ruled by their children—a world much relished by children. But to say that nature sanctions and supports as well as frustrates our lofty aspirations, that there may be "preëstablished harmonies" between its order and our ideals, is simply to restate the very premises of naturalism. Our values are the fine flowering of our purposes, and our purposes are rooted in natural conditions.

At any rate, the religious experience itself is primary. To have it in the name of God may no doubt facilitate, deepen, and enrich it; but the experience came first, before the Name. More important, it is not totally different in kind from other experiences, and it need not have a conventionally religious setting. It may be stimulated by a passage from Shakespeare, a moonlit ocean, a heroic or generous deed, or even simple comradeship and fellow feeling. Any experience that comes as an intense, unqualified good, that unifies the self and unites it with a larger, impersonal good, and that (unlike the momentary excitement of a football game) reinforces our most cherished goods, is religious in quality. As Dewey has said, the artist, scientist, parent, or plain citizen who is moved by "the spirit of his calling" is controlled by an unseen power; any working union of the ideal and the actual, any activity pursued at the risk of personal sacrifice because of a conviction of its enduring value, realizes the function usually attached to the idea of God.

There remains, however, the crucial question: whether such natural idealism, a religion without God, can have the necessary authority, force, and vitality to realize its values. The final argument for supernaturalism is always that it cannot. "Humanity will have the courage and enthusiasm to survive," Roberts asserts, "only if it believes that it is fighting for something more than its own survival." One cannot say positively that he is wrong —no one can collect bets on the survival of the race. Yet neither can Roberts be so positive. Biology does not support the notion that man alone of animals must be insured in the heavenly mutual before he will put up with life, nor does the evidence of history. If men have rarely worshiped Man, they have often worshiped men; they have had the courage and enthusiasm to fight to the death for worldly causes; and today the Communists are evidence enough that a secular religion can command as absolute faith, evoke as strong passion, produce as willing martyrs—inspire as extreme fanaticism as the Church. Supernaturalists too often think of disbelief as mere lack of belief, an aching void—which it often is. But it may also be a strong faith, positive in its belief in purely humanistic values. There is also a great tradition of inquiry and dissent, as ancient, dignified, and significant as the orthodox Christian tradition.

I should not pretend to give an unqualified answer to this crucial question. Obviously it is helpful to believe that humanity is playing for more than earthly stakes. The lives of many men have been steadied and ennobled by this belief; few men are so manly that they would actually prefer to get along without God. Nevertheless the terms of our problem are set by what men can believe, not what they would like to believe, and today faith can scarcely be so simple and certain as once it was. It would at least seem reasonable to explore the possibilities of getting along without God, to try to make the best of the conditions of our knowledge and experience. And at least a humanistic faith has certain real advantages.

For our knowings, it is clearly apt to be more fruitful. No new data can be inferred or foretold from the supernatural hypothesis, nothing added to what we have already learned from observation and experience in the natural world. If this hypothesis does prepare the mind for mystery (which is a real intellectual as well as moral need in this positivistic age), it also encourages mystery mongering; and in the long run we can always be sure of having plenty of mystery on hand, our greater need is always to dispel mystery. For our doings, therefore, a humanistic faith can also be more fruitful. It can retain the poetry and wisdom of Christianity, as of the other world religions; but because it conceives the "soul" of man as an achievement, not something he is born with and must merely "save," it must attend to the natural, social conditions of this achievement. It may cut away the irrelevant, outworn dogma that confuses religious issues; it may emancipate natural idealism from too rigid or narrow ideals; it may more accurately locate the means and ends of effective idealism. "Were men and women actuated throughout the length and breadth of human relations with the faith and ardor that have at times marked historic religions," Dewey declares, "the consequences would be incalculable."

Needless to add, this prospect is no more than a pious hope. Humanistic faith and ardor are scattered among different ideals; there is no catholic creed in this kind, and on naturalistic grounds too we can expect to have many religions. Yet the issues of conduct are not for that reason hopelessly confused, any more than they are by the controversies among all the sects. The revelations and commandments of orthodox Christianity are at times dubious and certainly often unclear; the immense literature of theology is a tacit confession that the will of God needs considerable interpretation. As for the God of the modernists or the mathematicians, He gives no real revelations or commandments at all; He merely gives our values a metaphysical nimbus. In specifying or urging their values, however, men perforce appeal to experience on this earth. Except for the most barren theological controversies, the final appeal is always to some kind of naturalistic standard.

Whether or not a man finds a faith without God emotionally as well as intellectually satisfying is meanwhile a matter of temperament and training. Meanwhile, too, society at large has suffered from the decay of traditional faith; it needs the equivalent of religion and has been the more unstable because the constituted spiritual authorities have been losing their hold on the citizenry. But even this serious problem becomes less alarming in a historical perspective. Intellectuals who worry about the masses forget that there has never been a truly Christian society, not even in the Middle Ages that Christian poets and philosophers now picture as the Golden Age; and that when the masses have been fired with religious zeal, they have usually expressed their zeal by persecuting rival sects, burning heretics, and starting religious wars. Religion has been a powerful social force, but not quite the kind of force that literary men describe.

Finally, we might reconsider the relations between religion and art. Their traditional intimacy is natural enough, since cultures tend to integrate their major activities. By the same token, however, it is not something special, and as we have seen, anthropology gives evidence that it is not essential. In our own culture their relations have manifestly grown more casual ever since the Renaissance. Assuming Whitehead's four main stages in the evolution of religion—ritual, emotion, belief, and rationalization—art plays an important role in the first two, and can still be helpful in the third; but in the last stage it is naturally considered superfluous, even dangerous. Protestants have often viewed it with suspicion or alarm. Moreover, art would appear to be more necessary to a vital faith than religion—an official creed—is to a vital art. In honoring the great cathedrals, we should remember not only that religion has also produced a great deal of dreary, mechanical art, but that there are many other masterpieces it did not inspire. Even grant Eliot's assertion that Shakespeare could have written still greater poetry had he had a Christian philosophy, he managed well enough without it.

But very possibly he managed better. In so far as Shakespeare's imaginative thought can be called a philosophy, it was broadly humanistic, and its great advantage was precisely the scope and freedom it permitted his genius. Like communism or any specific creed, Christianity may vitalize but also may fetter art—*Paradise Lost* is the obvious example of both the kindling power and the dead weight of dogma. It may limit the range of the artist's interests and felt responses. It may, indeed, restrict the possibilities of *religious* experience in art. For the important thing is that art in its own right gives direct access to the "wider self."

Ransom's "God without thunder," religion conceived as metaphor and myth, a symbolic means to appropriate attitudes—this is essentially poetry. Religion merely as poetry is not, once more, the religion of old; historically, Christianity has meant considerably more (and something less); but it remains great poetry after its claims to literal truth have been rejected. In any event, intense esthetic experience is one of the varieties of religious experience. Natural pieties are natural continuities, and by its rendering of the basic rhythms and patterns, art immediately realizes these continuities. By such natural means it leads us to a "deeper reality," gets us in touch with "unseen powers," gives us "intimations of immortality"—provides the kind of "spiritual" experience that can steady us in confusion, support us in failure, deepen the significance of all our activities and our values. That art can offer such satisfactions is one reason, in fact, why religionists often distrust it. If poetry reconciles one to an imperfect world, writes G. Rostrevor Hamilton, "weariness will not 'toss him' to the breast of God." This worry about a dearth of dissatisfaction and weariness is a rather curious reflection of the orthodox means and ends of deity. Nevertheless art remains perhaps the clearest example of the union of the material and the ideal that may be considered the sign of effective religious values.

In its ethical bearings, accordingly, literature has not been merely a copybook or a marginal illustration of the religious text. It is also a critic and a creator of morals. It is as diverse in its precept as religion, its disciples are as inconsistent in their practice; yet it is always pertinent. The sensitive writer is more likely to see around dogma and cult than is the professional moralist, and to keep an eye on the given terms, biological and social, of the moral problem. He finds morality where it is, in all our activity, not in a separate realm of value. He can see through the stock characters of the ethical drama, "innate good" and "innate evil": he naturally perceives the mixed stuff in man and the endless complication of circumstance, the conditions under which men *make* good and evil by their own efforts, with reference to their own purposes, and according to their own standards. He therefore keeps morality under constant criticism, helps to rescue it from legalism or the lower levels of reward and punishment. And he is the more effective because he usually works indirectly and unconsciously. Imagination is still, as Shelley said, the chief instrument of moral good; we always need "to imagine that which we know," and that which we will. It is the means to a vivid realization of the community of man, the immense sum of human striving—the social ties and needs that are the source of all morality.

The poet and dramatist naturally tend, moreover, to supplement the Christian ideal. They do not repress impulse and passion; they seek to clarify and direct them, to liberate them from mean ends and destructive means. Even as embodied in the Sermon on the Mount, the Christian ethic is still one-sided, incomplete; too strait is this gate, too narrow this way unto life. The increasing responsibilities of social and public life today alone demand a less passive, unworldly, ascetic morality, one that will give more prominence to purely secular duties, the salvation of society as well as of the individual soul. But at all times there is need to qualify the blessedness of the poor in spirit. Self-respect is healthier than self-contempt, courageous endeavor is more admirable than meek submission, high-minded independence is nobler than innocence. If it is all too easy to find instances now for the traditional sermon that Pride is the deadliest of the Seven Deadly Sins, Pride nevertheless remains a moral necessity, the spring of all personal dignity and responsibility. And there are also instances enough today of an actual poverty of spirit: Christian intellectuals who have become too aloof from the wickedness of this world, or grown too much at home with sorrow, or given themselves up too utterly to the conventional religious spirit, which has been allied more often with sadness than with strength or vigor or joy in life.

Nor does the state of the world today clinch the argument for the necessity of God. For two thousand years Christianity has failed to prevent terrible wars—when it has not itself generated them. Now, as always, God is on both sides; now, as always (except in religious wars), thoughts of God are chiefly afterthoughts as men fight out their all too worldly differences. In so far as they ardently believe in their cause they have, indeed, an essentially religious spirit; but they are proving that this spirit can be had without benefit of clergy. The present fervor for democracy is further evidence that a vital faith does not require supernatural sanctions. As for the ultimate question of the "cause" or "meaning" of the whole universe, the thorough-going naturalist can answer only that its order probably bears some analogy to human intelligence; and there is no doubt, as Hume observed, some reason for melancholy because this intelligence "can give no solution more satisfactory with regard to so extraordinary and magnificent a question." But Hume also observed that such panegyrical words as "infinite" and "eternal" are no real solution either, and finally as unnecessary for our moral as for our philosophical purposes. "Wise," "noble," "just," "magnanimous"—such words are big enough to fill the imaginations of men. For a long time to come, certainly, they will be lofty enough even for those whose reach must exceed their grasp, and for whom only the unattainable can be the true ideal.

4. THE RELATION TO DEMOCRACY

ALTHOUGH economic and political problems are outside the scope of this study, the humanist is forced by his premises to face the broad issues involved in them. The loftiest manifestations of a culture are never independent of its material conditions; the good society cannot exist without a sound basis in practical life. Specifically, the supreme issue of our time is the issue of democracy, and whether it is fit to survive. Few thinkers need to be told that they are vitally concerned in this matter; most have openly committed themselves. Yet there remains considerable ambiguity. Many radicals and conservatives alike have not really decided how far they are ready to go with the people; their respect for democracy depends upon the latest election returns. Similarly literary men veer between the extremes of a contempt for the Masses and a glorification of the People. Often they proclaim at once the dignity of Man and the immitigable vulgarity of common men; they despair of a democratic culture without being willing to commit themselves to an undemocratic government. They want freedom for everything, and they want to get it for nothing.

I hold that democracy, broadly conceived, is the necessary political faith of humanism, the necessary basis of the good society. So once more it will be well to begin by saying the worst about it. Like so many admirable ideals, it is supported by an unsound idea, the equality of all men, and it is prolific of illusions. Auguste Comte expected that the glorious society of the future would be built jointly by philosophers and the working classes; these "natural allies" resemble one another "in generosity of feeling, in wise unconcern for material prospects, and in indifference to worldly grandeur." If this portrait of the working classes has a quaint nineteenth-century air, that of many reformers today is scarcely more to the life. Nor is this a harmless idealization. Those who want democracy to work can least afford to overrate the wisdom and vigor of the common man. It is a pity that those who know him best are typically the advertisers, the politicians, the birds of prey.

Like religion, democracy must be considered as it is found; and its political life will often be found wanting. The opinions of the man on the street necessarily have great weight, even an aura of sanctity when they add up to a majority. Yet these opinions are a motley of hand-me-downs, baggy generalities and shabby prejudices picked up from the nearest bargain counter, with little eye for fit, harmony, or the occasion. Most people, Bertrand Russell once said, would sooner die than think—and in fact do so. Certainly the common man has little capacity for independent, disciplined thought, or when he has the capacity he still lacks the necessary knowledge and training. With the increasing complexity of our civilization, moreover, his limitations become increasingly dangerous. Even slight disturbances may throw so delicate a mechanism seriously out of gear; democracy can no longer afford so many of the demagogues and mediocrities he is apt to elect to high places. Neither can it indefinitely afford the cynicism reflected in the connotations of the word "politician." Karl Mannheim points out that democracy has also bred morally destructive influences. The robber morality once regarded as valid only in extreme cases and only for rulers is now more generally acknowledged and condoned; the cynicism of sophisticates becomes public property.

Nevertheless the masses continue to have only a superficial understanding of their civilization. Although they usually admire it, they live on credit, they live on a cultural dole, and what they admire is chiefly the gadgetry. In urging the noble ideal of transmitting our cultural heritage to all men alike, Lester Ward declared that all, even the feeblest, could easily support it: "The truth is no harder to carry than was the error; in many ways it is the lighter load." But this is to conceive culture and truth as mere baggage. They are instead an arduous discipline, a whole way of thinking and feeling, and any educator can testify how stubbornly the young idea is like to resist them. Meanwhile the masses are not only indifferent but often hostile to their rightful heritage. Although they stand in awe of the scientist, as a kind of witch doctor, they do not cherish the spirit of pure science or the ideal of free inquiry; they have a natural scorn of the "impractical," a natural resentment or fear of new ideas. Nor are they completely devoted to the democratic ideals that have brought about their own rise to power. The ordinary American cheers all mention of freedom and resents all interference with his own; he also keeps a jaundiced eye on his unconventional neighbor, joins the Ku Klux Klan, wants to throw all radicals into jail, and in his exalted moments becomes a Vigilante. The chief victims of his zeal for defending American ideals are precisely the most zealous champions of civil liberties or of his own welfare. Few tasks are more arduous and unrewarding than that of protecting him against himself. Hence Nazism itself is a genuine idol of the masses. It would be comforting to view Hitler as a tyrant fetched up by frightened capitalists; but he is so dangerous, monstrous, appalling because he embodies the hopes and fears of the mass mind on its lower levels.

That the masses have little passion for sweetness and light is of course nothing new in history. What is new, however, is the ubiquity of their own preferences in culture. All our mighty machinery is in one aspect simply a means of amplifying their voice and multiplying the signs of their taste. Our amazing instruments of entertainment and edification produce the high-powered vulgarity of the movies, the radio, the tabloid. Our immense "institutions of higher learning" struggle to teach the rudiments of learning, turning out hordes of fancy illiterates and of technicians whose command of one subject fortifies their ignorance and scorn of all others. And the total growth, rank and gross though it is, is also subtly contaminating, for it pervades all areas of activity, scatters seeds at a distance, leads to unconscious compromises. When a society is dedicated to the proposition that every man is as good as every other man, and that all alike are theoretically entitled to the same privileges regardless of their fitness for exercising them, the natural tendency is to reduce all standards to the lowest common denominator.

This is to point to the distinction between common men and uncommon men. To make the distinction is embarrassing: by inference one ranges himself with the elite. Yet in a democracy, above all, the integrity of an elite is important for culture. As it is, the fear of being snobbish or effete leads thinkers to make a virtue of the very limitations of the common man. They talk as if a man were necessarily more "alive" when working in field or factory than when in laboratory or library, and as if his values were then necessarily more "real," more deeply and richly human than the values of philosopher, scientist, or artist—just as wealthy men often talk of the blessedness of poverty. Much realism in modern literature is but one illustration of this indiscriminate democratic ideal. Writers seem to consider all subjects equally deserving of treatment, interminably recording the "reactions" of mean or trivial heroes who have no reactions to speak of. In various ways they minimize the differences between men—the differences that may be slight to one contemplating their common mortality, but remain the source of all civilized standards.

Altogether, it is easy to draw up a formidable indictment of the sovereignty of the people, with the incessant high-powered manifestations of low-powered minds that are in one view the norm of modern civilization. Yet this is a one-sided view. Much complaint of democracy is itself vulgar, an expression of class prejudice implicitly assuming that the upper classes in society as now organized are inherently superior. Much is also sentimental, reflecting a nostalgia for some great age when culture was in flower, with a convenient forgetfulness of the great masses who then as now did the dirty work, and whose tastes and interests were not improved by this work. But almost all the familiar criticism is greatly oversimplified; and here is the case for the defense.

To begin, the "masses" are in fact masses of men—not a single force but a vast heterogeneity of interests and loyalties. Snobbishness or intolerance imposes the unity, creates the monster, just as does the totalitarian state. To be sure, the mass is not a mere aggregate, it notoriously has its own psychology; but it can intensify the loyalties and idealisms as well as the fears and hatreds of its members, tap the mysterious sources of heroism of men united in a common quest and a common hope. ("There's more nobility in the *ensemble* of my squadron," wrote Malraux in *Man's Hope,* "than in almost any of the individuals composing it.") In general, the mass man is of very mixed stuff and not merely fearsome. If he is too easily taken in by demagogues or fanatics, he can also be commanded by honorable men, fired by noble causes, and he is capable of a simple dignity, fidelity, and fortitude that often shame his leaders. If he has made Nazism his own, he is also bearing the brunt of the war against it—and in the established democracies is more trustworthy than many of his betters. Even as a voter he compares favorably enough with them. Business and professional men can usually be counted on to vote the Republican ticket—scarcely on high intellectual grounds; the prejudices of the disinherited would seem no more deplorable or dangerous than the inveterate self-interest of the privileged.

Meanwhile, it is easy to exaggerate the influence of the common man. Politically, his power is often nominal, his freedom always in danger of becoming a more ingenious or complicated form of slavery; there is more reason for despairing of the realization of the democratic ideal than for worrying about its effects upon standards. As for culture, the facts simply do not support the notion that this ideal has been a fatal blight. Here again literary men are prone to a romantic, sentimental view of the past, in which a Dante is seen as the true type of his age and the sweating, brawling masses are not seen at all. There is no room in our materialistic society for "such a monster as the cultivated man," wrote Delmore Schwartz (for an audience of such monsters); the poet, he added, cannot exercise his culture and sensibility in writing about the lives of his fellow men because in them "culture and sensibility [have] no organic function." Actually, his fellow men have never been distinguished for their sensibility; yet the greater poets have managed to exercise their own—often, like Chaucer, writing about their cruder fellows, and sometimes, like Shakespeare, even writing for them. And so can the poet today, as the greater novelists do.

We can do without the common proofs of our improving culture. That 277,794 Americans attended an Italian Masters exhibit, or even that 12,000,000 of them listen to "Information, Please," leaves 100,000,000 unaccounted for, and the question of appreciation untouched. Nevertheless there remains plenty of respectable cultural activity, plenty of opportunity for devotees of the most unpopular kinds of beauty and truth. The pure sciences are clearly prosperous as never before; there appear to be easy chairs enough for historians, philosophers, and scholars generally; there is no dearth even of fine art. As for commercial relations with publishers and dealers, they are no more undignified than the patronage of a noble lord, and usually far more profitable. The danger, indeed, lies in the too easy rewards: men sell out their talents, or are content merely to repeat a popular formula. Nevertheless the bitch-goddess Success is a very ancient deity. Artists and thinkers today need make no sacrifice of integrity, and the best of them have made none. I doubt that any genius, or many robust, first-rate talents, have been either frustrated or devoured by the masses.

On the contrary, the rise of the masses has plainly liberated much talent and expanded opportunity. "It takes all kinds to make a world," says the common man. If he does not fully appreciate the humanistic truth of this truism, if he is often suspicious or intolerant of any but his own kind, he is still useful for much more than the hewing and hauling. A society of intellectual aristocrats has its own limitations and dangers. Lancelot Hogben has noted the failure of the brilliant Greeks to make noteworthy use of geometry; freed from the vulgarity of practical use, it became a hobby of intellectuals, in Plato a means to "spiritual perfection"—and so it ended where it began, in superstition. Furthermore, common men can provide all the necessary kinds to make a world. A knowledge of the workings of heredity would relieve the widespread distress today over the failure of the "superior" classes to reproduce themselves. The standard of superiority is not always clear; but if the favored few, by whatever standard, produced no offspring whatsoever, the great masses would supply society with almost as many people of their type. And a democracy, finally, is more apt to discover and use these many exceptional children of the poor. Years ago John M. Robertson demonstrated that genius is by no means sure to work its way to the front. Of some hundred of the most famous European writers between 1265 and 1865, only two—Bunyan and Burns—came from the lowest class. We can be sure that today much superior talent still runs to waste; despite our cherished success stories, the bright newsboy is still only a thousand to one shot to become a captain of industry. But we can also be sure that society now salvages much more.

Similarly we may ease somewhat the worry over related conditions of our culture. Ever since Mill, men have complained that individuality is being crushed under the despotism of public opinion and standardized custom; yet in the same period customs have been revolutionized, all fields of thought and activity have been stirring with more experiment and innovation than ever before, and the arts in particular have been conspicuous for a riot of individuality and nonconformity. And so with the Machine that is to make robots of us all. Modern technology has indeed brought much ugliness, destroyed many spiritual landmarks, compelled an oppressive standardization and regimentation. It has also brought immense imaginative exercises in the world of affairs, the grandeur and the glory realized in concrete and steel, the great bridges, dams, trains, planes, liners. The human cost seems greater because old forms of drudgery are romanticized, old forms of slavery forgotten; it is doubtful whether the peasants of the past were greater individualists than factory hands today, or their work less soul-killing. Altogether, beneath the all too obvious and depressing uniformities, modern society is far more various than any before. It has many more frontiers, offers many more opportunities for the cultivation of unusual interests and abilities.

In general, we may too easily overlook, because we take for granted, the remarkable abundance that has come with the growth and spread of the democratic ideal. No era in history has been more imaginative, original, adventurous, or creative than the modern era; no world empire of the past has had more intoxicating visions than the kingdoms of common men today. The powers and possibilities of man have been enormously increased, his outlook enormously expanded. If the quality of our immense quantities is often dubious, and the ends of our immense means are even more so, the possibilities are still there and the achievements still stand. It is idle to welcome them, yet deplore the conditions that made them possible. It is idle to ignore the necessity of the masses, or to regard them simply as a necessary evil. It is imperative to make the best of democracy —unless one can propose a feasible alternative. And aside from the totalitarians, the critics of democracy usually propose none.

There is no sure-fire solution to the practical problems of democracy: no way of guaranteeing an intelligent delegation of power, an efficient and wise administration of power. But neither are there guarantees in any other kind of state. Although the ideal government presumably would be the administration of a trained, enlightened aristocracy, or of a great and good dictator, wholly dedicated to the best interests of the whole citizenry, no one has yet suggested an adequate means of instituting, securing, and perpetuating such a government: a means of getting around the notorious facts that blooded aristocracies are prone to decadence as well as to selfishness, that wise and strong kings or dictators are like as not to be succeeded by imbeciles or weaklings, that no ruling class has ever failed to abuse unlimited power. Even Nazism, one may venture safely if not cheerfully, is ill equipped for the long run. On its historical record, meanwhile, democracy seems at least as satisfactory as other forms of government. In theory, it is still the most reasonable form. Among the most admirable and unobjectionable of our common sentiments is that every man ought to be given a fair chance. If it has been far from perfectly realized in the democracies, it nevertheless points to the more serious objections to undemocratic governments. The common man is the chief victim of social inefficiency and injustice, the first to pay for the mistakes of his superiors; and the ballot, however abused, is the most practicable check upon these mistakes.

If one goes this far, he has then to accept the cultural consequences. A responsible concern for the values of culture demands an intelligent regard for their actual basis and frame. The political duties of the common man logically require his education, he could not in any event be given the ballot and then be put in cultural quarantine; but the main forces in the modern world are automatically working toward a diffusion of knowledge and culture. Human society, once more, is distinguished from animal societies primarily by its development of conscious communication; it is a community of meanings, an association in spiritual as well as physical goods. Modern society is distinguished from earlier societies primarily by its immense extension and improvement of the means of communication; lacking our facilities, the great societies of the past had more limited possibilities because they were necessarily more mechanical, rigid structures. Science, which has made our civilization possible, is akin to democracy in its faith in human intelligence and coöperative effort; in its own community it recognizes no distinction of class or race, insists upon free inquiry and free expression. If it tends to lose touch with the whole community, its more elementary findings and its by-products nevertheless become to an increasing extent the property of all men. Thus it has incidentally supplied the technological means for making works of art universally available; and the social value of the arts lies in their contribution to a rich and moving communication, breaking down the barriers between men. In short, that all the members of a society possess its culture, share in its heritage, would seem to be a natural ideal; but it is at any rate a natural tendency for our kind of society.

We can accept this tendency as a fact without applauding all the fetishes that pass for education and culture. We need actually to *make* the best of democracy, indeed; and this effort calls for not only patience and generosity but a clear, steady idea of the best. It calls for an uncompromising attack on the kind of democratic sentimentality reflected in Tolstoy's doctrine that all art should be comprehensible to the common man. Most men are now patently unable to appreciate the acknowledged masterpieces, for the same reason that they are unable to understand Einstein; even in a classless society it would still be absurd to expect ordinary minds fully to appreciate the greatest achievements of the greatest minds. At best there inevitably are different levels of response. The simple reader who enjoys Shakespeare —or Tolstoy himself—is not having the same experience as a trained, sensitive reader. Like it or not, in short, most great art is the possession of a minority; to deny this is simply to depreciate art—and without improving the condition of the common man, who will continue to know what he likes. But grant this and I can see no possible objection to the democratic effort to raise the level of response, to enlarge the minority, above all to make the minority a wholly elective and not a hereditary body.

Hence there is need of stressing again the contributions of the great majority. If literature is chiefly the property, it has not been the exclusive creation of an elite. It has often been refreshed and revitalized from popular sources, has always drawn from folkways and folklore. Folk art, in fact, is the oldest and strongest tradition in art (though generally slighted by critics like Eliot who make a specialty of Tradition), and is a valuable reminder of the homely, natural, communal function of art. Especially valuable today, moreover, are the resources of popular speech. Granted that its vigor and raciness are overrated, it has contributed more than accredited literature to the problem of naturalizing our new world, finding new symbols for our old needs. Abstractions have been multiplying far more rapidly than concrete terms and metaphors, a flood of technical terms carries toward the semantic ideal of mere naming without feeling; but the people have been steadily renaming, inventing the necessary metaphors, bringing the abstractions back to life, reclaiming the technical in homely terms, finding poetic symbols in the new environment—finding "beauties" long before most artists realized that poetry was not necessarily tied to the symbols, beliefs, and myths of tradition. They have been doing unconsciously, anonymously, consistently what Hart Crane tried to do in *The Bridge*. "Everything comes out of the dirt—everything," said Walt Whitman: "everything comes out of the people, the everyday people, the people as you find them and leave them . . . just people!" Speak this instead of chanting it, add the necessary reservations, pay the proper tribute to the great creative genius, and there still remains this truth: that deep roots can be sunk only in the dirt, and that no art or no thought can long afford to be out of touch with the people.

Indeed, more unfortunate than the vulgarization of culture is the persistence of the aristocratic tradition. The aloofness from science and industry, practical knowledge and common use, and all the material conditions of modern society is a hang-over from a world that no longer exists. However charming a world, it cannot be brought back; and even those who are unhappy over its disappearance are seldom willing to assume the responsibility of its necessary economic and political conditions. "Since we can neither beg nor borrow a culture without betraying both it and ourselves," writes Dewey, "nothing remains save to produce one." We must produce one under the given conditions of our kind of society, attempt to realize the full potential significance of our kind of life. The serious objection to the cultural isolationists like Schwartz is that they side-step the whole problem, finally aggravate the problem by accentuating both the crudities of an unenlightened industrial life and the superficialities of an ungrounded cultural life. At worst, they encourage snob values, which are not so very different from mob values; the esthete is much like the vulgarian in his fondness for conspicuous virtuosity or novelty, the external signs of culture, and he confirms the vulgar conception of the artist as a kind of cultural beautician.

At best, the isolationists maintain the old forms without the old substance; and this is the sign of shabby gentility.

All this is not to say that the artist must become one of the boys, directly enlist in some popular or useful cause. It is only to repeat that in the long run he must be in some vital relation to the public world, even if the relation be a passionate protest, and cannot afford to be simply disdainful of the slovenly unhandsome mob betwixt the wind and his nobility. And if the high-pressured materialism and mediocrity of the public world explains the tendency to fastidious withdrawal, democracy also provides the antidote. Its natural tendency—and in literature, on the whole, its actual tendency—is to promote a vigorous, responsive, and responsible art. From Milton to Mann, most writers have not been oppressed by its conditions; many have plainly found in democratic ideals a source of inspiration and strength.

In the end, Hitler is the most obviously effective critic of democracy; he has brutally simplified and forced the issue. Yet he has also complicated the issue, even apart from raising genteel misgivings about the propriety of combatting the wave of the future. His criticism could not in any event be met simply by a wishful preference in futures, but he forces a careful statement of faith. I believe that democracy is also a positive force, and that revolution and reaction are not the only forms of action; I stand on Hegel's famous pronouncement, that history is the history of liberty. Still, the definitions here are all-important. In one sense totalitarianism is the extreme expression of the principle of individualism to which the democracies have been devoted. All significant achievement is "the exclusive achievement of the individual," Hitler has said; all government must therefore insure "freedom from the principle of control of majorities—the masses—in order to achieve the undisputed authority of the Individual." He has achieved this kind of freedom, he exercises this authority; and by democratic standards it means slavery and tyranny. Nevertheless he has also moved in the direction of state socialism, a planned economy, a world order—the direction that more and more men of good will have come to believe is necessary. He has exploded the favorite argument of conservatives, that socialism may be all right in theory but it won't work; his planned economy has proved far more efficient than free private enterprise. And so the idea is current that the ideal of individual freedom is doomed by the inevitable trend toward collectivism.

Now I incline to agree that some sort of planned society is inevitable, at least if modern civilization is to survive. It is required by the intricate interdependencies of this kind of civilization; all the leading states have been moving toward more government control, even under the Hoovers and the Baldwins, for they have been forced despite themselves by the increasing pressure of problems that the most rugged individuals could not handle. Likewise some sort of world order is a logical, if not inevitable outcome. In the long view, the evolution in social behavior has been an irregular, intermittent, but demonstrable tendency to merge or transcend local interests, to extend rights and duties to ever larger groups, to widen the sphere of loyalties from the family to the community, the community to the tribe, the tribe to the nation—with the more restricted loyalties not eliminated but included in a larger whole. If such an evolution is no guarantee that the next steps will be successfully taken, from the nation to the league of nations, from the league to the world, it makes this a reasonable goal. Isolationism is an understandable sentiment, but it no longer corresponds to the social facts.

Yet implicit in this conception of internationalism is the objection to a world order imposed by a "master race," maintained by force and not by loyalty or mutual interest. Such an order is not organic, and on the face of it cannot be stable. I also believe —and here is a more difficult issue—that a planned society not only can but should be democratic. Efficient planning calls for the preservation of our ancient liberties if only because it must be experimental, flexible, self-critical, and will need our full resources of intelligence. The planners must be controlled because they cannot themselves be perfectly planned. It will indeed be a hazardous experiment; I cannot know whether men will command the necessary intelligence and good will to make it work. But meanwhile men are needlessly distressed by the actual trend in government because they identify individualism with laissez-faire economics, liberty with nineteenth-century liberalism, as if the principle of democracy had a fixed form and content. The problem is not to preserve these forms at all costs but to redefine and readapt the basic principle. This has been the aim of the more thoughtful New Dealers, for example, however questionable their accomplishment. Similarly Mannheim attempts, by a concrete sociological analysis, to determine just what sort of democracy and individual liberty would be possible in a planned society.

It is idle to argue for freedom and against regimentation in the abstract; everybody wants freedom, nobody wants regimentation per se. It is essential to have a clear idea of our actual freedom, and then to ask the all-important question: freedom for *what?* On the whole, the increasing "regimentation" in the democracies has meant more real freedom for more men, more opportunity for a fuller development of individuality, and in a better planned society there conceivably could be far more still. As for the ideal, it is not to be free for everything and from everything. For man, who cannot trust to instinct or act on every impulse, it must be a rational ideal. In the words of Charles H. Cooley, it is "freedom to be disciplined in as rational a manner as you are fit for." The possibilities accordingly vary with individuals as with whole societies; some are fit for more than others. And the immediate question is whether the citizenry of the advanced democracies will prove fit for the degree of freedom that they have had.

At the end, we do well to return to the simplicities. Far back in 1938, critics were demanding referents for such words as "democracy" and calling meaningless such statements as "Fascism is a menace to civilization." Today we need not be told that these words are precisely the kind most pregnant with meanings. and that a bomb sight can locate them well enough for practical purposes. We feel that the willingness of men to fight for an idea may be no less conclusive proof of its reality than certification by a logical positivist. Far back in 1940, many men were still of two minds and half a heart. "There has seldom been a generation," Mannheim wrote, "which was less willing for petty sacrifice and more likely to pay the supreme one without even understanding why." Today we are still more likely, but we better understand why. We mind such ideas as justice and liberty; we know that they matter. Altogether, I hold that the democratic ideals evolved by the growth toward conscious culture are essential to the continued development and even the continued existence of this culture. If they are incompatible with the very choicest kind of culture, the most exquisite flowers of the human spirit, then there is no lasting soil for these flowers and we must do without them. The ideals are their own sufficient justification, and the more necessary to preserve because they have been so imperfectly realized in the world's democracies.

5. CODA

A HORSE, D. H. Lawrence sadly observed, is always true to its pattern; but a man you can never rely on. You cannot indeed. As a higher animal, man has no set pattern; he is free to choose, and to be false to his choice. Neither can you rely on nature to take care of her favored child. Nature is taking more of a chance with him than with her other species; because his behavior is not guaranteed by instinct, he runs greater risks. If you stress the wonderful potentialities of man, you must admit the worst possibilities, and if you stress the wonderful plasticity of nature, you must add that it is also adaptive to his most stupid or vicious purposes. At that you must confess, as James did in his old age, that for the purposes of a lifetime reality isn't so damn plastic after all. In any event, there can be no perfect rest or peace for any living organism. A measure of instability, imbalance, disharmony is the very condition of life, because implicit in the fact of growth and change. The consciousness of man but intensifies the endless striving for an equilibrium and security that can never be attained. And if you say that contemporary man has the especial misfortune of being a wanderer between two worlds, you must add that science dooms him to a perpetual wandering; it demands that thought live dangerously, and it leaves him always a prey to the

> Blank misgivings of a creature
> Moving about in worlds not realized.

In short, the thoroughgoing naturalist cannot easily sustain the high note appropriate to a Conclusion, even if the times were more propitious for such exaltedness, and a philosophy now were not chiefly useful, perhaps, as an aid to endurance. I end as I began, in the provisional and uncertain.

Yet the provisional and uncertain remain the sign of possibility, freedom of thought and action, and because the naturalist may readily declare the conditions, he may also subordinate the main clause of these melancholy statements. You cannot rely on man—but least of all as you respect him; if he were as true as the horse he would not be its master. If there are obvious reasons for complaint of nature's arrangements, there could be no consciousness and no purpose at all in an utterly congenial environment because no function for thought, no significant choices to make. Tension, resistance, conflict, insecurity are the source of all our interests and our values; the passing is the means to the surpassing, the token of what we most prize in experience. We like it to be all one, but only at the end. Good-in-evil, the permanency of the transitional, the constancy of the contingent—these are the irrevocable terms of human life. To the humanist they are honorable terms, because the necessary condition of the human spirit. The sign of spirit is doubt, disquiet, endless seeking and not finding; this is also its dignity. Almost all the great writers have told this same story, and told it, like Mann today, with irony and with reverence.

In the long view, the dual view, we can see why the great spirits have typically been wanderers, the great ages unstable ages, and why, as Whitehead has said, it is the business of the future to be dangerous. Today, when the future is doing an especially conscientious job, men may take to this view simply as a refuge from the immediate prospect. Yet it is still a necessary perspective, for men are more inclined to see our world simple and sell it short. All great social achievements are also complications, creating further problems; all societies have the defects of their virtues, their overemphases and their unintended byproducts; all prove the observation of William James, that without too much we cannot have enough, of anything. Men used to dwell on the striking achievements of our civilization, the distinctive virtues, the abundance, the positive element in the inner contradictions. Now they dwell on the complications, the defects, the excess, the self-destructive element. They are prone to reduce multiple possibilities to a single inevitability; literary men in particular like to contemplate the doomed and the damned, because nothing is more dramatic. And though intellectuals can scarcely intensify the violence that has already come to be, they can confuse the issues of what ought to be, what still could be.

The ordinary citizen today knows as little about the social forces that regulate his life as the savage knew about the forces of nature, and is therefore as easy prey to blind hopes and fears, as apt to put his trust in magical solutions. The intellectual knows more, but he is seldom qualified to speak with authority on political, economic, and technological problems, the effective means of controlling social forces; so he too is susceptible to the simple or wishful solution, recommending some ideal society—agrarian, classless, Christian—without regard for the effective means. Today he is directly involved in tremendous issues, which have to be fought out whether or not they can be thought out. Nevertheless his primary responsibility, as an intellectual, is to think them out. He should at least be more farsighted than the untrained citizen, more critical, enterprising, and verbally resourceful, able to give under the pressure of events without losing his consistency and sense of direction. As a public-spirited citizen he may passionately commit himself, but as an intellectual he also needs the long view, the dual view. As it is, many intellectuals have been winning cheap victories, which are always likely to lead to ignominious defeats. And others have been proclaiming the loss of a battle as the failure of the whole campaign.

Thus Allen Tate sees in the humanism of Matthew Arnold the symbol of our failure. "His program, culture added to science and perhaps correcting it, has been our program for nearly a century, and it has not worked." Nearly a century!—when Christianity is still on trial after two thousand years. And in all history, what program has worked? Worked "for good," and wholly for good? More to the point, have we actually given Arnold's program a trial? As for science, it is easy to say that we have too much: more than we can assimilate, more than we can manage. It is as easy to add that we need still more, in order to manage what we have; the sciences of man have only begun, the scientific spirit has scarcely entered the administration of society. As for culture, at any rate, the humanism of Arnold has no more governed our intellectual life than it has informed our political life.

T. S. Eliot, however, takes a much longer view of our experiment. In "East Coker," his recent testament, he raises the first and last issues; and on his terms, finally, I am content to declare my own testament. "In my beginning is my end," runs the refrain of his poem. The beginning was the Renaissance: the faith in the power of intelligence and knowledge, which has created the modern world.* This faith is a delusion, Eliot declares:

* For the necessary exegesis I am indebted to James Johnson Sweeney. Few readers could have picked up by themselves the clue in the title: Coker is the reputed birthplace of Sir Thomas Elyot (d. 1564), a Renaissance humanist who was also classicist, royalist, and Anglo-Catholic.

There is, it seems to us,
At best, only a limited value
In the knowledge derived from experience.
The knowledge imposes a pattern, and falsifies,
For the pattern is new every moment.

Right now the lights are being extinguished, the darkness is closing in; and the poem (Sweeney writes) "celebrates the culmination of suffering and purgation." But Eliot also anticipates the Resurrection. Through the only wisdom, the "wisdom of humility," we may now return to "the eternal source of truth" to which he appealed in *The Idea of a Christian Society*. It is the "intuition of pure being" that Christopher Dawson calls the starting point of human progress, because it leads to the knowledge of Divine Being. And so the poem comes to a triumphant conclusion:

In my end is my beginning.

Now there is some reason to doubt not only the toughness but the thorough humility of a spirit that cannot endure evil unless it is called Original Sin. Humility may be a little cheap on these terms; the celebration of suffering seems too easy when the faithful are promised a happy ending. Nevertheless Eliot's thought and feeling go deep, he has not come by them easily, and in any event he forces the fundamental issue. The knowledge derived from experience is in fact limited, its patterns must forever change; still I hold by it as the only knowledge we can be sure of. Through the ages men have gone to some eternal source of truth, and had intuitions of pure being; still their truths and their intuitions look very different, and I have heard of no way to decide which are the eternal and pure ones. The medieval world, whose end was our beginning, was presumably closer to the idea of a Christian society; still it ended, presumably for good reasons, and I conclude that the knowledge of Divine Being also has a limited value, at least as the foundation of a society. If all that Eliot says may be so, on the face of the evidence it does not look so. I stick to the evidence.

It would be foolish to deny the possibility that our civilization will collapse. History is big enough to swallow us too. Yet if the experiment of intelligence does fail, I still believe that it was worth trying, that it was not doomed from the beginning, that it might have had a better end—and that, like the experiments that preceded it, it will leave much more than some picturesque archaeological ruins. If the lights are now going out, I affirm that they were lights, and that we are then indeed going into the darkness. Hitler represents something very old in history; he too is an absolutist, his mythology is also an intuition. Meanwhile we can still make history. We need the heart and the mind to do it, we need more than patience and prayer. Eliot tells us to be still and breathe deep:

I said to my Soul, be still, and wait without hope
For hope would be hope for the wrong thing; wait without love
For love would be love of the wrong thing; there is yet faith
But the faith and the love and the hope are all in the waiting.
Wait without thought, for you are not ready for thought:
So the darkness shall be the light, and the stillness the dancing.

They also serve; but I prefer to trust, hope, and love even at the risk of choosing the wrong things. I believe that only by trying to think can one get ready for thought. I prefer faith with works.

INDEX

The following is primarily an index of names. It also contains, however, a number of references to important themes whose development is not adequately indicated by the Table of Contents.

THE YALE PAPERBOUNDS

Y–1 Liberal Education and the Democratic Ideal *by A. Whitney Griswold*
Y–2 A Touch of the Poet *by Eugene O'Neill*
Y–3 The Folklore of Capitalism *by Thurman Arnold*
Y–4 The Lower Depths and Other Plays *by Maxim Gorky*
Y–5 The Heavenly City of the Eighteenth-Century Philosophers *by Carl Becker*
Y–6 Lorca *by Roy Campbell*
Y–7 The American Mind *by Henry Steele Commager*
Y–8 God and Philosophy *by Etienne Gilson*
Y–9 Sartre *by Iris Murdoch*
Y–10 An Introduction to the Philosophy of Law *by Roscoe Pound*
Y–11 The Courage to Be *by Paul Tillich*
Y–12 Psychoanalysis and Religion *by Erich Fromm*
Y–13 Bone Thoughts *by George Starbuck*
Y–14 Psychology and Religion *by C. G. Jung*
Y–15 Education at the Crossroads *by Jacques Maritain*
Y–16 Legends of Hawaii *by Padraic Colum*
Y–17 An Introduction to Linguistic Science *by E. H. Sturtevant*
Y–18 A Common Faith *by John Dewey*
Y–19 Ethics and Language *by Charles L. Stevenson*
Y–20 Becoming *by Gordon W. Allport*
Y–21 The Nature of the Judicial Process *by Benjamin N. Cardozo*
Y–22 Passive Resistance in South Africa *by Leo Kuper*
Y–23 The Meaning of Evolution *by George Gaylord Simpson*
Y–24 Pinckney's Treaty *by Samuel Flagg Bemis*
Y–25 Tragic Themes in Western Literature *edited by Cleanth Brooks*
Y–26 Three Studies in Modern French Literature *by J. M. Cocking, Enid Starkie, and Martin Jarrett-Kerr*
Y–27 Way to Wisdom *by Karl Jaspers*
Y–28 Daily Life in Ancient Rome *by Jérôme Carcopino*
Y–29 The Christian Idea of Education *edited by Edmund Fuller*
Y–30 Friar Felix at Large *by H. F. M. Prescott*
Y–31 The Court and the Castle *by Rebecca West*
Y–32 Science and Common Sense *by James B. Conant*
Y–33 The Myth of the State *by Ernst Cassirer*

Y–34 FRUSTRATION AND AGGRESSION *by John Dollard et al.*
Y–35 THE INTEGRATIVE ACTION OF THE NERVOUS SYSTEM *by Sir Charles Sherrington*
Y–36 TOWARD A MATURE FAITH *by Erwin R. Goodenough*
Y–37 NATHANIEL HAWTHORNE *by Randall Stewart*
Y–38 POEMS *by Alan Dugan*
Y–39 GOLD AND THE DOLLAR CRISIS *by Robert Triffin*
Y–40 THE STRATEGY OF ECONOMIC DEVELOPMENT *by Albert O. Hirschman*
Y–41 THE LONELY CROWD *by David Riesman*
Y–42 LIFE OF THE PAST *by George Gaylord Simpson*
Y–43 A HISTORY OF RUSSIA *by George Vernadsky*
Y–44 THE COLONIAL BACKGROUND OF THE AMERICAN REVOLUTION *by Charles M. Andrews*
Y–45 THE FAMILY OF GOD *by W. Lloyd Warner*
Y–46 THE MAKING OF THE MIDDLE AGES *by R. W. Southern*
Y–47 THE DYNAMICS OF CULTURE CHANGE *by Bronislaw Malinowski*
Y–48 ELEMENTARY PARTICLES *by Enrico Fermi*
Y–49 SWEDEN: THE MIDDLE WAY *by Marquis W. Childs*
Y–50 JONATHAN DICKINSON'S JOURNAL *edited by Evangeline Walker Andrews and Charles McLean Andrews*
Y–51 MODERN FRENCH THEATRE *by Jacques Guicharnaud*
Y–52 AN ESSAY ON MAN *by Ernst Cassirer*
Y–53 THE FRAMING OF THE CONSTITUTION OF THE UNITED STATES *by Max Farrand*
Y–54 JOURNEY TO AMERICA *by Alexis de Tocqueville*
Y–55 THE HIGHER LEARNING IN AMERICA *by Robert M. Hutchins*
Y–56 THE VISION OF TRAGEDY *by Richard B. Sewall*
Y–57 MY EYES HAVE A COLD NOSE *by Hector Chevigny*
Y–58 CHILD TRAINING AND PERSONALITY *by John W. M. Whiting and Irvin L. Child*
Y–59 RECEPTORS AND SENSORY PERCEPTION *by Ragnar Granit*
Y–60 VIEWS OF JEOPARDY *by Jack Gilbert*
Y–61 LONG DAY'S JOURNEY INTO NIGHT *by Eugene O'Neill*
Y–62 JAY'S TREATY *by Samuel Flagg Bemis*
Y–63 SHAKESPEARE: A BIOGRAPHICAL HANDBOOK *by Gerald Eades Bentley*
Y–64 THE POETRY OF MEDITATION *by Louis L. Martz*
Y–65 SOCIAL LEARNING AND IMITATION *by Neal E. Miller and John Dollard*
Y–66 LINCOLN AND HIS PARTY IN THE SECESSION CRISIS *by David M. Potter*

Y-67 SCIENCE SINCE BABYLON *by Derek J. de Solla Price*
Y-68 PLANNING FOR FREEDOM *by Eugene V. Rostow*
Y-69 BUREAUCRACY *by Ludwig von Mises*
Y-70 JOSIAH WILLARD GIBBS *by Lynde Phelps Wheeler*
Y-71 HOW TO BE FIT *by Robert Kiphuth*
Y-72 YANKEE CITY *by W. Lloyd Warner*
Y-73 WHO GOVERNS? *by Robert A. Dahl*
Y-74 THE SOVEREIGN PREROGATIVE *by Eugene V. Rostow*
Y-75 THE PSYCHOLOGY OF C. G. JUNG *by Jolande Jacobi*
Y-76 COMMUNICATION AND PERSUASION *by Carl I. Hovland, Irving L. Janis, and Harold H. Kelley*
Y-77 IDEOLOGICAL DIFFERENCES AND WORLD ORDER *edited by F. S. C. Northrop*
Y-78 THE ECONOMICS OF LABOR *by E. H. Phelps Brown*
Y-79 FOREIGN TRADE AND THE NATIONAL ECONOMY *by Charles P. Kindleberger*
Y-80 VOLPONE *edited by Alvin B. Kernan*
Y-81 TWO EARLY TUDOR LIVES *edited by Richard S. Sylvester and Davis P. Harding*
Y-82 DIMENSIONAL ANALYSIS *by P. W. Bridgman*
Y-83 ORIENTAL DESPOTISM *by Karl A. Wittfogel*
Y-84 THE COMPUTER AND THE BRAIN *by John von Neumann*
Y-85 MANHATTAN PASTURES *by Sandra Hochman*
Y-86 CONCEPTS OF CRITICISM *by René Wellek*
Y-87 THE HIDDEN GOD *by Cleanth Brooks*
Y-88 THE GROWTH OF THE LAW *by Benjamin N. Cardozo*
Y-89 THE DEVELOPMENT OF CONSTITUTIONAL GUARANTEES OF LIBERTY *by Roscoe Pound*
Y-90 POWER AND SOCIETY *by Harold D. Lasswell and Abraham Kaplan*
Y-91 JOYCE AND AQUINAS *by William T. Noon, S.J.*
Y-92 HENRY ADAMS: SCIENTIFIC HISTORIAN *by William Jordy*
Y-93 THE PROSE STYLE OF SAMUEL JOHNSON *by William K. Wimsatt, Jr.*
Y-94 BEYOND THE WELFARE STATE *by Gunnar Myrdal*
Y-95 THE POEMS OF EDWARD TAYLOR *edited by Donald E. Stanford*
Y-96 ORTEGA Y GASSET *by José Ferrater Mora*
Y-97 NAPOLEON: FOR AND AGAINST *by Pieter Geyl*
Y-98 THE MEANING OF GOD IN HUMAN EXPERIENCE *by William Ernest Hocking*
Y-99 THE VICTORIAN FRAME OF MIND *by Walter E. Houghton*
Y-100 POLITICS, PERSONALITY, AND NATION BUILDING *by Lucian W. Pye*

Y–101 More Stately Mansions *by Eugene O'Neill*
Y–102 Modern Democracy *by Carl L. Becker*
Y–103 The American Federal Executive *by W. Lloyd Warner, Paul P. Van Riper, Norman H. Martin, Orvis F. Collins*
Y–104 Poems 2 *by Alan Dugan*
Y–105 Open Vistas *by Henry Margenau*
Y–106 Bartholomew Fair *edited by Eugene M. Waith*
Y–107 Lectures on Modern Idealism *by Josiah Royce*
Y–108 Shakespeare's Stage *by A. M. Nagler*
Y–109 The Divine Relativity *by Charles Hartshorne*
Y–110 Behavior Theory and Conditioning *by Kenneth W. Spence*
Y–111 The Breaking of the Day *by Peter Davison*
Y–112 The Growth of Scientific Ideas *by W. P. D. Wightman*
Y–113 The Pastoral Art of Robert Frost *by John F. Lynen*
Y–114 State and Law: Soviet and Yugoslav Theory *by Ivo Lapenna*
Y–115 Fundamental Legal Conceptions *by Wesley Newcomb Hohfeld*

THE YALE WESTERN AMERICANA PAPERBOUNDS

YW–1 Trail to California *edited by David M. Potter*
YW–2 By Cheyenne Campfires *by George Bird Grinnell*
YW–3 Down the Santa Fe Trail and Into Mexico *by Susan Shelby Magoffin*
YW–4 A Canyon Voyage *by Frederick S. Dellenbaugh*
YW–5 An Introduction to the Study of Southwestern Archaeology *by Alfred Vincent Kidder*
YW–6 The Fur Trade in Canada *by Harold A. Innis*
YW–7 The Truth about Geronimo *by Britton Davis*
YW–8 Sun Chief *edited by Leo W. Simmons*
YW–9 Six Years With The Texas Rangers *by James B. Gillett*
YW–10 The Cruise of The Portsmouth, 1845–1847 *by Joseph T. Downey*